Tribal GIS

Supporting Native American Decision Making

EDITORS

Anne Taylor
David Gadsden
Joseph J. Kerski
Heather Warren

Esri Press
REDLANDS | CALIFORNIA

Cover art sources: Map courtesy of Chickasaw Nation; image from Shutterstock, courtesy of Jeffrey M. Frank

Esri Press, 380 New York Street, Redlands, California 92373-8100

Copyright © 2012 Esri
All rights reserved.
Printed in the United States of America

16 15 14 13 12   1 2 3 4 5 6 7 8 9 10

Library of Congress Cataloging-in-Publication Data
Tribal GIS : supporting Native American decision making / Anne Taylor ... [et al.].
   p. cm.
   Includes bibliographical references.
   ISBN 978-1-58948-320-0 (pbk.)
1. Geographic information systems—United States. 2. Decision making--United States. 3. Indians of North America—Information services—United States. 4. Indians of North America—Effect of technological innovations on—United States. I. Taylor, Anne, 1961–
   G70.215.U6T75 2012
   910.285—dc23
                                                                                                                                        2012001144

The information contained in this document is the exclusive property of Esri unless otherwise noted. This work is protected under United States copyright law and the copyright laws of the given countries of origin and applicable international laws, treaties, and/or conventions. No part of this work may be reproduced or transmitted in any form or by any means, electronic or mechanical, including photocopying or recording, or by any information storage or retrieval system, except as expressly permitted in writing by Esri. All requests should be sent to Attention: Contracts and Legal Services Manager, Esri, 380 New York Street, Redlands, California 92373-8100 USA.

The information contained in this document is subject to change without notice.

U.S. Government Restricted/Limited Rights: Any software, documentation, and/or data delivered hereunder is subject to the terms of the License Agreement. The commercial license rights in the License Agreement strictly govern Licensee's use, reproduction, or disclosure of the software, data, and documentation. In no event shall the US Government acquire greater than RESTRICTED/LIMITED RIGHTS. At a minimum, use, duplication, or disclosure by the US Government is subject to restrictions as set forth in FAR §52.227-14 Alternates I, II, and III (DEC 2007); FAR §52.227-19(b) (DEC 2007) and/or FAR §12.211/12.212 (Commercial Technical Data/Computer Software); and DFARS §252.227-7015 (DEC 2011) (Technical Data – Commercial Items) and/or DFARS §227.7202 (Commercial Computer Software and Commercial Computer Software Documentation), as applicable. Contractor/Manufacturer is Esri, 380 New York Street, Redlands, CA 92373-8100, USA.

@esri.com, 3D Analyst, ACORN, Address Coder, ADF, AML, ArcAtlas, ArcCAD, ArcCatalog, ArcCOGO, ArcData, ArcDoc, ArcEdit, ArcEditor, ArcEurope, ArcExplorer, ArcExpress, ArcGIS, ArcGlobe, ArcGrid, ArcIMS, ARC/INFO, ArcInfo, ArcInfo Librarian, ArcLessons, ArcLocation, ArcLogistics, ArcMap, ArcNetwork, *ArcNews*, ArcObjects, ArcOpen, ArcPad, ArcPlot, ArcPress, ArcPy, ArcReader, ArcScan, ArcScene, ArcSchool, ArcScripts, ArcSDE, ArcSdl, ArcSketch, ArcStorm, ArcSurvey, ArcTIN, ArcToolbox, ArcTools, ArcUSA, *ArcUser*, ArcView, ArcVoyager, *ArcWatch*, ArcWeb, ArcWorld, ArcXML, Atlas GIS, AtlasWare, Avenue, BAO, Business Analyst, Business Analyst Online, BusinessMAP, CityEngine, CommunityInfo, Database Integrator, DBI Kit, EDN, Esri, Esri—Team GIS, Esri—The GIS Company, Esri—The GIS People, Esri—The GIS Software Leader, FormEdit, GeoCollector, Geographic Design System, Geography Matters, Geography Network, GIS by Esri, GIS Day, GISData Server, GIS for Everyone, JTX, MapIt, Maplex, MapObjects, MapStudio, ModelBuilder, MOLE, MPS—Atlas, PLTS, Rent-a-Tech, SDE, SML, Sourcebook·America, SpatiaLABS, Spatial Database Engine, StreetMap, Tapestry, the ARC/INFO logo, the ArcGIS logo, the ArcGIS Explorer logo, the ArcPad logo, the Esri globe logo, the Esri Press logo, the GIS Day logo, the MapIt logo, The Geographic Advantage, The Geographic Approach, The World's Leading Desktop GIS, *Water Writes*, arcgis.com, esri.com, geographynetwork.com, gis.com, gisday.com, and Your Personal Geographic Information System are trademarks, service marks, or registered marks of Esri in the United States, the European Community, or certain other jurisdictions. CityEngine is a registered trademark of Procedural AG and is distributed under license by Esri. Other companies and products or services mentioned herein may be trademarks, service marks, or registered marks of their respective mark owners.

Ask for Esri Press titles at your local bookstore or order by calling 800-447-9778, or shop online at esri.com/esripress. Outside the United States, contact your local Esri distributor or shop online at eurospanbookstore.com/esri.

Esri Press titles are distributed to the trade by the following:

In North America:

Ingram Publisher Services
Toll-free telephone: 800-648-3104
Toll-free fax: 800-838-1149
E-mail: customerservice@ingrampublisherservices.com

In the United Kingdom, Europe, Middle East and Africa, Asia, and Australia:

Eurospan Group
3 Henrietta Street
London WC2E 8LU
United Kingdom
Telephone: 44(0) 1767 604972
Fax: 44(0) 1767 601640
E-mail: eurospan@turpin-distribution.com

Contents

Foreword by Jack Dangermond vii
Preface ix
Map of contributors x

Chapter 1 : Managing tribal resources and protecting the environment   1

Chapter 2 : Moving ahead in transportation   13

Chapter 3 : Preserving tribal culture and history   25

Chapter 4 : Economic development on tribal lands   49

Chapter 5 : Creating healthy communities   67

Chapter 6 : Connecting communities through primary, secondary, and informal education   87

Chapter 7 : Building career pathways through higher education   107

Chapter 8 : Fostering sustainable tribal agriculture and rangelands   127

Chapter 9 : Ensuring tribal safety   135

Chapter 10: Supporting the tribal enterprise   143

Foreword

As some of the earliest adopters of geographic information system (GIS) technology, Native American tribal governments have used GIS to support thousands of programs and initiatives. Today, Tribes use GIS throughout their organizations to address their unique challenges as sovereign Nations. GIS has become a common platform for problem solving by helping Tribes organize and analyze data as well as collaborate and communicate on countless issues. Among these issues are natural resource management, transportation, cultural and historic preservation, realty, economic development, health, education, agriculture, and public safety.

This book offers insight into how tribal governments and supporting organizations are employing GIS, from day-to-day operations to special projects for tribal leadership. *Tribal GIS: Supporting Native American Decision Making* also highlights how GIS is being used to embrace a new movement in tribal governance toward improving citizen services, decision support for tribal leadership, sustained economic development, and the protection of tribal assets.

Increasingly, Tribes are realizing that geographic data provides an integrating framework for all tribal government activities. As GIS has evolved into an enterprise platform for information management, many Tribes are establishing enterprise GIS platforms within their governments, enabling them to better address the complex challenges of sovereign Nations.

I hope you enjoy reading about how Tribes across the United States are using GIS, and that you will continue to explore the important and innovative ways that geographic knowledge can be used and shared today and in the future.

Warm regards,

Jack Dangermond

Jack Dangermond
President, Esri

Preface

Tribal GIS: Supporting Native American Decision Making provides Native policymakers, administrators, scientists, and instructors with information on how GIS and the spatial perspective can be used to make their organizations more efficient and more effective. It illustrates how GIS is used to solve problems on tribal lands and in tribal programs on a local, regional, and national level. Although other documents have been published on GIS use in Native American communities, none have been told by the tribal leaders themselves. Their stories critically assess not only the benefits of GIS, but the challenges they had to overcome for successful implementation. The stories clearly indicate that GIS is most effective when implemented enterprise-wide, which benefits multiple departments and has long-lasting benefits.

Each chapter is organized around a major theme, including natural resources and the environment, transportation, cultural and historical preservation, economic development, health, education, public safety, agriculture, and enterprise GIS. Each chapter includes several stories illustrating the chapter's theme, sometimes at a local scale, sometimes on a regional or national level, but always focused on the spatial perspective. They explain the difference GIS makes on the efficiency of programs, the lives of people, and the health of the planet.

While reading these stories, consider the following questions, which are relevant to all GIS projects:
- How was spatial data gathered, managed, and updated?
- What issues benefited from spatial analysis?
- How did the quality of the data affect the analysis and conclusions of the projects?
- How were sensitive tribal records and cultural resource locations protected and secured?
- How did the use of GIS affect the project's size, management, and workflows?
- Were the results of the spatial analysis presented effectively to different stakeholders?
- What GIS skills were needed to perform the analysis?
- What important skills should students develop for careers in GIS?

This book was written by those who overcame organizational, technological, and financial obstacles to implement something they believed would improve the quality of life for everyone. The goal in compiling these stories is to help those who haven't yet implemented a GIS to consider the advantages that GIS can bring to an organization. It is also hoped that those who are already working with GIS can learn from the practices implemented by those who have blazed the trails.

Acknowledgments

We are deeply grateful to each of the authors who generously and willingly shared their stories in this book. We thank all those who provided us with permission to use the images contained in these pages. We also thank Kirsten Grish for contributing her rich photographs featured throughout the book, and Tim Clark for creating the map of all the Native places represented in the book. We wish to express our gratitude to Esri Press, for its support of this project, and to Heather Warren for her tremendous assistance that came at a critical time. Lastly, we salute the GIS practitioners on tribal lands who are daily making wise decisions that are making this world a better place.

Map of contributors

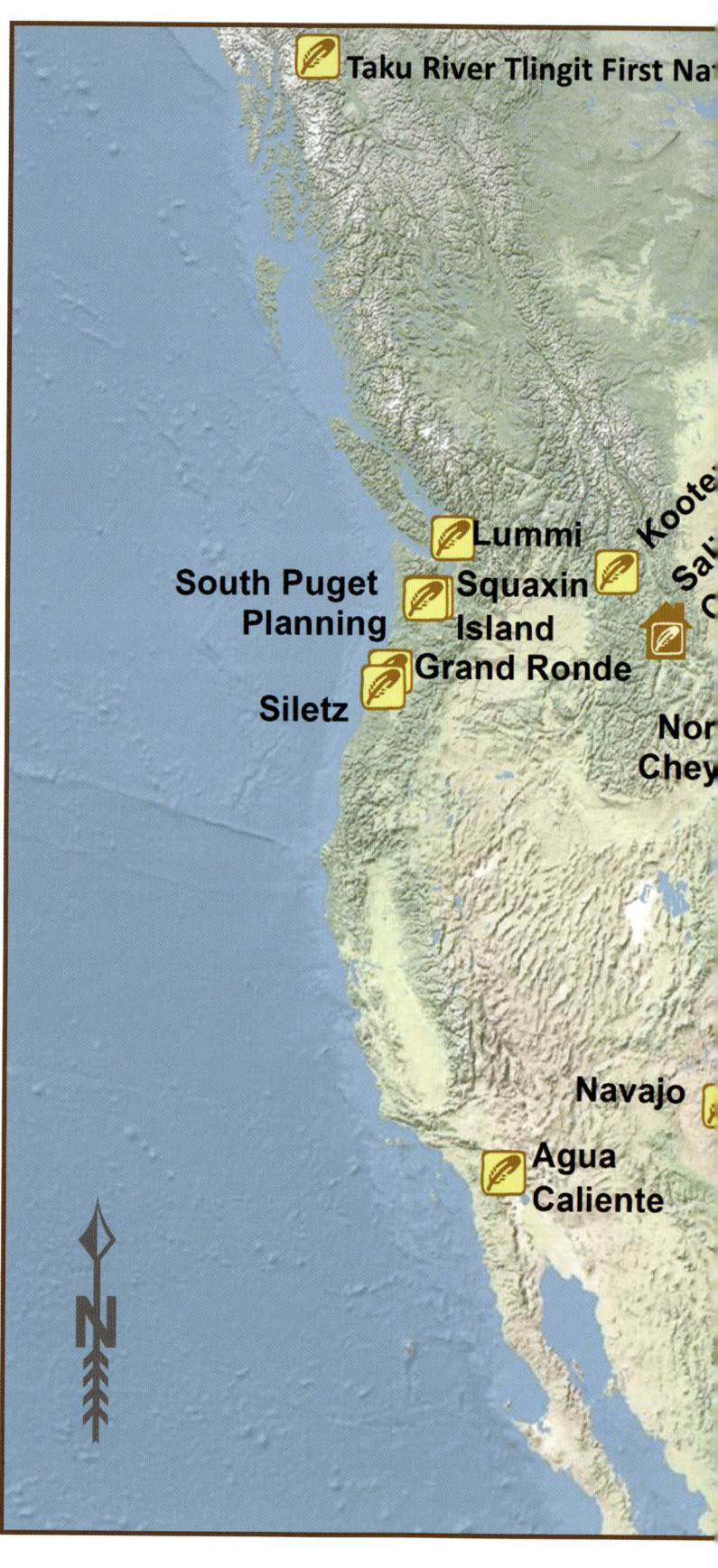

Photo by Joseph Kerski

Managing tribal resources and protecting the environment

Native American tribal governments are some of Esri's original customers. Tribal forestry, range, and fisheries programs have used GIS for several decades now. The extensive use of GIS by tribal governments in support of natural resource and environmental management should not come as a surprise. Native Americans have strong personal, cultural, and spiritual ties with the earth. A Tribe's cultural identity is inextricably tied to the places they and their ancestors have lived and worshiped. The animals and plants that inhabit tribal areas represent a natural living fabric over these lands, which provide the food, shelter, and medicines that rejuvenate and sustain Native American People. There are many individuals and unique threads, or assets, which contribute to this fabric. These assets can be abstracted in a GIS, allowing them to be better understood in the context of complex, ever-changing landscapes.

By embracing GIS technology, a Tribe can conduct a detailed inventory of current natural resources as well as incorporate traditional ecological knowledge on that resource's quality and health generations back. Traditional knowledge and information technology are not mutually exclusive. By incorporating the wisdom and experiences of tribal Elders into a common framework with current observations on natural resources, environmental trends become obvious.

As the relationships and dependencies within ecosystems are better understood, the state of tribal natural resources can be projected. GIS can spatially model dynamic relationships, allowing natural resource management practices to be virtually tested and rated on their reflection of the Tribes' values and long-term objectives. The health and abundance of the community's natural environment is a major factor in the health of the tribal community. GIS provides an integrated approach to visualizing these complex environments and analyzing the relationships within them. Through spatial analysis, unique models can be developed to help set priorities, suggest appropriate harvesting methods, and assist in developing long-term sustainable management practices. Tribal lands are not infinite, and Tribes are increasingly challenged to maintain a limited footprint. This situation often requires tribal lands to be used in multiple ways; however, ill-advised land-use changes are likely to impact other tribal interests. Therefore, a multi-dimensional framework, such as a GIS, is required to provide visibility of these complex relationships and dependencies. GIS provides a platform for integrating diverse layers of information into a common operational view where the relationships across natural resource and environmental assets provide important context. GIS allows critical areas to be mapped, managed, and protected so that scarce and sacred resources can be maintained for future generations while economic development opportunities are maximized.

GIS professionals from three tribal governments have contributed to this chapter on the use of GIS for the management of natural resources and the environment; they include the Confederated Tribes of Siletz Indians, Squaxin Island Tribe, and Confederated Tribes of Grand Ronde, with usual and accustomed areas reaching across the US states of Oregon and Washington. Each Tribe has applied GIS to unique management challenges and has shared their stories and approaches here for the benefit of other tribal and indigenous communities.

Using GIS to protect critical wildlife habitat

Brady Smith, GIS Planner; and Mari Kramer, Assistant Tribal Forester
Confederated Tribes of Siletz Indians
Siletz, Oregon

The Siletz Tribe is a confederation of a number of western Oregon Tribes on one reservation. Prior to treaties and removal, individual villages were autonomous and each one acted politically independent. Tribal members took only what was needed and respected the land, fish, and wildlife. With ample territory for their subsistence needs, villages and populations did not infringe upon each other's hunting areas. These principles allowed for the Siletz ancestors to call western Oregon home for thousands of years.

In fulfillment of treaty promises, western Oregon Tribes were moved to a 1.1-million-acre Coast (Siletz) Reservation established in 1855. However, most of the reservation was opened to settlement by presidential and congressional actions in 1865, 1875, and 1892. In 1954, the Western Oregon Termination Act was passed, ending federal recognition of the Siletz Tribe. After many years of complex negotiations, the Siletz Tribe was restored to federal recognition in 1977. In 1980, the Tribe had 3,630 acres of original reservation timberland and thirty-six acres known as Government Hill returned as reservation land in Siletz, Oregon. Following restoration, the Siletz have worked diligently to recover as much of their ancestral lands as possible. Today, the Siletz Tribe has managed to acquire additional acres of timberland and real estate, through purchase or gift, within the original reservation boundaries.

The Tribe realized several years ago that GIS could be an effective tool for land management. In the early 1990s, GIS was used primarily for tracking forest management activities. The GIS program has faced a variety of challenges over the years, including turnover of experienced personnel, which hindered the overall success of the program. After a number of years of not having a dedicated GIS person, the Tribe reevaluated the value of GIS, and a position with stable funding was created in 2004 that would serve all tribal programs. This resulted in a more centralized approach to GIS, allowing key information to be shared between tribal departments. Today, the Tribe's GIS platform provides decision support for the following tribal programs: Tribal Administration, Real Estate, Planning and Community Development, Fish and Wildlife Management, Forestry, Culture, and Education.

The Siletz People have always tried to protect land, fish, and wildlife for future generations. However, their environmental stewardship was dealt a severe challenge in 1999 when a ship, the M/V *New Carissa*, ran aground just north of the entrance to Coos Bay. Approximately 25,000 to 140,000 gallons of tar-like "bunker" fuel were lost to the marine environment. The impact was devastating, particularly on wildlife. Over 260 marbled murrelets, a threatened bird species, were killed in the event. As part of a settlement with the responsible party, approximately 4,260 acres of timberland were purchased to compensate the public for injuries to marbled murrelets caused by the spill. The land is located within the Siletz River Basin and consists of predominantly commercial timberlands.

The Siletz Tribe was selected as the land manager over two conservation groups, and the property was transferred with a conservation easement, restricting activities on the land to create habitat for the threatened species. As part of the conservation easement, the Tribe had to develop a management strategy that would protect habitat while generating enough revenue to maintain the land.

The first step was to develop a baseline report for the property. This initial report included information on the property's natural resources (timber, fish, wildlife, soils, streams, and vegetation), adjacent land ownerships, property boundary surveys, water rights, road use agreements, and other relevant management information. The information was collected from various sources, including timber cruises (sample measurements of sections of forests that are used to determine an estimate of the amount of standing timber volume that the forest contains), wildlife surveys, property line surveys, state and county records, and GPS data collected by the Tribe. An unforeseen obstacle in the collection of data, between the start of the project and the completed report, was a change in the property boundary as a result of a new land survey. This boundary change required the redevelopment of several GIS layers that were critical to the accurate analysis of the data. GIS was a valuable tool in developing the baseline report narratives, tables, and maps. It was also a key tool in developing the property management plan.

The desired outcome for the management of the property is to have areas that would serve as a habitat for marbled murrelet as well as areas that would generate revenue for property management activities. To implement this management strategy, a polygon layer with three protection area categories (Highest, Buffer, and Standard) and two sub categories each was developed using the baseline information. These protection areas dictate where different types of management activities will occur on the property. Areas designated as the Highest Protection Area (HPA) classification are occupied marbled murrelet habitat as well as unoccupied suitable habitat. Buffer Protection Areas (BPA) include buffers of

300 feet around the HPA to protect habitat areas from disturbance and for development of future habitat. Standard Protection Areas (SPA) are the remaining areas not classified as either HPA or BPA.

By using GIS to map the protection areas for the property forest, managers realized it would be difficult to manage all activities on the property. Some management activities would need to occur within habitat areas, whereas other higher impact activities would need to be limited to areas where there was no habitat. Following a review of the baseline report, a list of planned activities for the property was identified. In order to effectively model where these activities would be allowed, GIS overlays were used to guide appropriate zones for various activities, which were referred to as "overlay zones." Based on GIS analysis, a final strategy was presented to the conservation easement grantees and the Siletz Tribal Council in the form of a management plan. In addition to the documents developed in support of the final plan, GIS was used during the public outreach related to the program. Large-format maps were used to help inform the public about the plan. This helped to effectively convey the Tribe's management strategy to the public and to the conservation easement grantees.

GIS allows tribal natural resource managers to visualize, plan, and monitor future management activities while considering protection areas and overlay zones. It also helps ensure that the management strategy is consistent with stipulations outlined in the conservation easement by establishing the areas that will become habitat. Using GIS helped the managers understand what areas on the property would produce the revenue to manage the property and what areas would be managed as protected habitat.

Currently, the geographic information about this property is stored in a geodatabase managed by ArcGIS Server. Data is used in several different platforms, including ArcGIS Desktop, ArcReader, and ArcGIS Server. It has also been recently incorporated into a web map application, which allows for easy access to the data for non-GIS users. Without the use of GIS in this project, the development of an effective management strategy would have been difficult to accomplish in the relatively short timeline required. This investment in GIS has also led to new opportunities for the Tribe and its resources. The GIS system was leveraged to support a successful grant for habitat restoration on a critical fish spawning stream on the property. The project will provide habitat restoration for salmonids and lamprey eels along seven miles of the stream. A helicopter will be used to place nearly 500 logs, half with root wads, in the stream to improve the habitat for the fish. GIS supported the planning of this project and supported the identification of source trees, helicopter staging areas, and in-stream placement sites for the logs. In these examples, GIS provided both a framework to support tribal decision making and the consideration of various land management strategies, as well as an actionable tool to aid in habitat restoration activities.

Developing a tribal natural resource atlas

Levi Keesecker, Quantitative Services Manager, Natural Resources Department
Squaxin Island Tribe, Washington State

Members of the Squaxin Island Tribe are Native Peoples of Puget Sound, an inland waterway that travels far from the ocean through a mystical braid of channels and islands between the fortress of the Olympic

Mountain Range to the west and the Cascade Range to the east. Squaxin Island tribal members are descendants of maritime people who lived and prospered along the southern extent of the sound for untold centuries. Because of their strong cultural connection with the water, they are also known as the People of the Water: the *Noo-Seh-Chatl* of Henderson Inlet, *Steh Chass* of Budd Inlet, *Squi-Aitl* of Eld Inlet, *Sawamish/T'Peeksin* of Totten Inlet, *Sa-Heh-Wa-Mish* of Hammersley Inlet, *Squawksin* of Case Inlet and *S'Hotle-Ma-Mish* of Carr Inlet. Through a treaty with the US government, one small island, four-and-a-half miles long and a half-mile wide, was reserved as the main area for all of the Tribe to live. The island was named after the People of Case Inlet and became known as Squaxin Island. Tribal headquarters are now located in Kamilche, between Little Skookum and Totten Inlets, where today hundreds of acres of land have been purchased and a thriving tribal community has been established. Despite the many interests and priorities of the Squaxin Peoples, the natural landscape and its resources remain a primary focus for the Tribe. Since 2003, the Squaxin Island Tribe has been applying GIS technologies to manage fisheries, watersheds, streams, environmental monitoring, and to plan future development. Tribal resources are not unlimited and often-difficult decisions need to be made as to where the Tribe can make the most impact in improving sensitive habitats.

The annual return of salmon to tribal waters is a fundamental cultural and subsistence resource for Native Peoples of the Northern Pacific Coast. The family of fish generally referred to as "salmon," or salmonids, follow a complex life cycle that begins with juvenile fish starting out in streams and rivers and migrating out to the open ocean to mature. Once fully matured, the salmonids return to the same freshwater pools where they began their lives to spawn the next generation.

Figure 1. Mature salmon return to their freshwater spawning grounds. From Shutterstock, courtesy of Daniel W. Slocum

When the Squaxin Island Tribe set out to optimize the impact of its natural resource management programs for the benefit of salmon populations, it leveraged GIS to model the areas of highest impact and vulnerability of salmonids habitats. Tribes as well as local governments and nongovernmental organizations in South Puget Sound often have difficulty prioritizing near-shore conservation and restoration projects. This is due to the limited amount of funding for restoration projects, the large amount of near-shore habitat in the region, considerable variation in habitat quality, and a complex tapestry of land ownership along the shoreline. Using GIS, the Tribe developed a Juvenile Salmonid Near-Shore Project selection tool to aid local governments and nonprofit organizations in selecting areas for habitat restoration and conservation projects. This project was funded by a grant from the Washington State Salmon Recovery Funding Board.

The process began with a detailed requirements gathering phase, which identified what type of information product would be most effective given the needs and technical capacity of end users of the system. Gathering detailed requirements from the community that will eventually use the GIS tools is critical to successful GIS implementations. Thoughtful advance consideration of the users' needs is required to appropriately apply the capabilities of GIS technology to the challenge at hand. Taking the time to meet with potential users and to inform and engage tribal leaders not only allows the collection of important requirements for the GIS system design, but also obtains support and buy-in along the way. Giving users a voice in the design of the system allows them a sense of contribution and ensures that the system reflects the broad needs of the Tribe. When broad user requirements are obtained, the stakeholder community feels as though it contributed to the system design and the resulting analysis was accomplished in part through its efforts. Following the requirements process for the Squaxin Island Near-Shore Project selection tool, it was determined that the final product should be a simple-to-interpret map that clearly depicts areas of high conservation or restoration value. In order to determine where the high value areas were located, a methodology was adopted from a previous study in neighboring Mason County, Washington (Anchor Environmental 2004).

One of the powerful abilities of a GIS is incorporating a variety of data elements into a composite view of a given environment. It is one thing to independently map current land-use practices along the shoreline, land ownership, and areas where salmonids are known to be present. One approach taken by the Squaxin Island Tribe is to bring these and many other spatial data layers together to depict the environment in its totality by modeling the impact these layers have on a given outcome, in this case the habitat quality for salmonids. GIS modeling referred to as geoprocessing provides this capability, allowing multiple layers of spatial data to be considered together in a comprehensive view of the environment being modeled. Once a model is authored, it can be run over large areas limited only by the availability of the spatial data used by the model. In the case of the Squaxin Island geoprocessing model, the Mason County model was modified and scaled to allow the analysis of all of South Puget Sound with over 400 miles of shoreline. Using the ModelBuilder framework with ArcGIS Desktop, geoprocessing models can be designed, tested, and run.

Using the ModelBuilder framework, the Squaxin Island Juvenile Salmonid Near-Shore Project selection tool applies GIS modeling on a larger scale to prioritize the restoration benefit of parcels along southern Puget Sound, based on a significant number of input layers and a variety of tools that contribute to the final output. As in our simple examples shown here, GIS layers (blue rectangles) are coupled

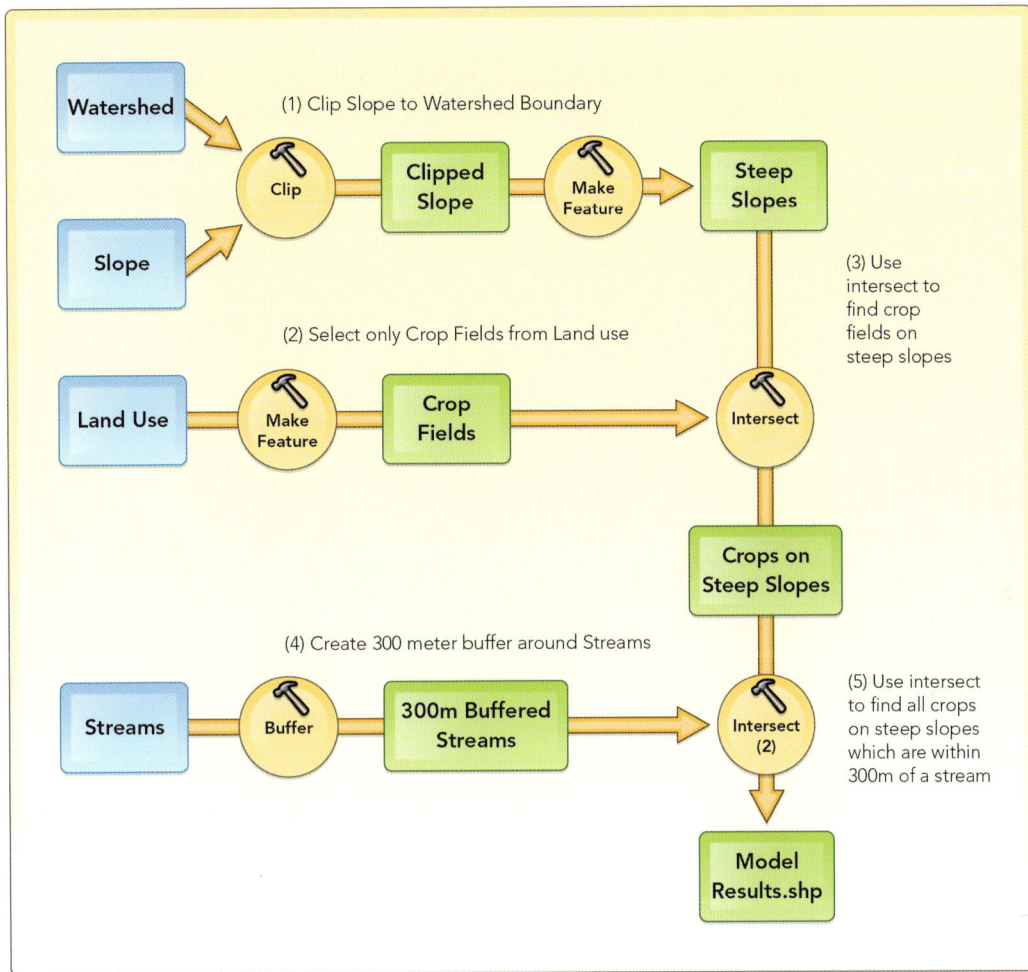

Figure 2. In this sample model, multiple criteria are considered to determine which crops on steep slopes are within 300 meters of a stream.

with a series of geoprocessing tools (orange circles) and output layers (green rectangles) to author a complex analysis. In this model, geoprocessing tools are used that assign a positive or negative "weight" to input layers based on their contribution to salmonid habitats. Weighting can be thought of as allowing each layer (or attributes within a layer) to have a vote in the final output of the model.

ModelBuilder also supports the nesting of one model within another. In this way, a model that initially excludes areas that are not relevant to the test can first be invoked to reduce the areas that are participating in the subsequent analysis. Other subroutines may iterate a model based on data that changes on a daily or monthly basis. In this way individual records in a table can be taken one at a time and run through the model, resulting in a mosaic dataset that shows change and trends over periods of time and over large geographic areas. The Juvenile Salmonid model is in fact a series of models that together provide a final composite result. Initially a set of three, limiting models are run, excluding areas that include features that would limit salmonid use, such as the presence of docks or overwater structures. Once these areas are excluded, a series of nine benefit models are run, which identify habitat features that would benefit salmonids, including proximity to a salmon-bearing stream, pocket estuary, presence of intertidal vegetation, or forage fish spawning beaches. This approach of building a beneficial and limiting model is that it allows you to determine areas that should be good for salmon (high benefit) that have high limiting scores, which would potentially be restoration sites. Areas with high benefit and low limiting are ideal conservation sites and are used to guide land acquisition activities by the Tribe and conservation trust organizations.

The results of the Juvenile Salmonid Near-Shore Project selection model identified the near-shore areas most important for conservation or restoration. As a result of this GIS analysis, future habitat restoration efforts are now prioritized based on a defensible consistent geographic analysis for South Puget Sound. In addition to benefiting the Tribe, this project benefits nonprofit organizations and local governments, as it allows them to target resources more efficiently. GIS provides a common framework for collaboration across a complex multi-jurisdictional landscape. Three local governments in the Puget Sound region have adopted the results of the model into their shoreline management plans as it represents the best available science on priority habitat areas for Puget Sound. Other nonprofit organizations have incorporated the model into their conservation programs. Oftentimes, landowners will contact these organizations for guidance on improving their land for the benefit of the environment. These organizations are able to use the GIS model output to locate the landowner's parcel and determine if it is an area of high or low restoration value in order to recommend the most appropriate conservation measures.

Increasingly, Tribes lead natural resource management negotiations across the private, governmental, and nonprofit stakeholders. The use of GIS is a powerful tool, allowing Tribes to inventory, track, and protect the health of critical resources. As Tribes are able to bring more land back into tribal jurisdiction, land acquisitions can be prioritized based on the highest return in habitat and natural resources for the benefit of the Tribe and surrounding communities. As Tribes adopt GIS technologies and make investments in key spatial data, staff, and technology, a new era of decision support and collaboration is emerging. GIS and spatial analysis can inform managers on how to apply limited program resources in a manner that ensures the highest gain. At Squaxin Island, GIS is used to better understand and protect threatened resources.

References

Anchor Environmental. 2004. Greater Mason County Near-Shore Habitat Assessment. Prepared by Anchor Environmental for Squaxin Island Tribe.

GIS for timber sale planning

Jason Bernards, Forester; and Volker Mell, GIS Coordinator
Confederated Tribes of Grand Ronde, Oregon

Since time immemorial, numerous Tribes lived in their traditional territories in western Oregon. Their territories extended from the Columbia River south of the Klamath River, and from the crest of the Coast Range to the crest of the Cascade Range. Beginning in 1853, these Tribes signed treaties with the US government, ceding away their traditional homelands for a permanent reservation in the Coast Range. In 1856, the Rogue River, Umpqua, Shasta, and Takelma Peoples were relocated from Table Rock and Umpqua Reserves to the Grand Ronde Reservation in northwestern Oregon. They joined the Kalapuya, Chinook, and Molalla Tribes and Bands from the Willamette Valley and the Columbia River that joined the reservation from the east.

In 1954, the Grand Ronde Reservation was terminated by the federal government. Termination was a devastating federal policy that severed the relationship between the Tribes and the federal government, resulting in the loss of federal services and benefits to the tribal members. In 1983, the federally recognized status of the Confederated Tribes of Grand Ronde was restored, and the Reservation Act of 1988 restored 9,811 acres to the Tribe, which is located between Salem and the northern coast of Oregon in the Coastal Mountain Range. The Tribe is now a sovereign Nation with the ability to protect its distinct differences, ways, and customs as an indigenous People while providing governmental services to the tribal members. The Tribe's vision is to be a community of caring people, dedicated to the principles of honesty and integrity, building community, responsibility, and self-sufficiency. The Tribe promotes personal empowerment and responsible stewardship of human and natural resources and strives to be a community willing to act with courage in preserving tribal cultures and traditions for all future generations.

The natural resources of the 9,811 acres of forested land managed by the Grande Ronde Natural Resources Department (NRD) provide timber production, fish and wildlife habitat, and recreational use. The Grand Ronde NRD uses GIS for day-to-day decision support on forest management, wildlife biology, and environmental protection issues. Since the natural landscape is constantly changing, field NRD staff collect GIS data on a regular basis, along with their other primary activities. This constant field data collection when incorporated into the NRD GIS platform allows finite lands under tribal jurisdictions to be managed effectively for multiple uses with the highest return. One benefit of this platform is support for efficient timber planning and harvest activities while protecting sensitive wildlife areas.

The Grande Ronde NRD program has taken an enterprise GIS approach, which makes spatial data and services available to all tribal government programs through the centrally managed GIS department. In this model, the geodatabase is used to store GIS information from different tribal government offices in a secure manner. In the 1980s and early 1990s, GIS layers were traditionally stored in file formats such as an ArcInfo coverage or a shapefile. Using GIS layers stored as files becomes problematic when multiple individuals need access to the same information. There is also no support for security or compression or a simplified way to back up a large number of file-based GIS layers.

In the mid-1990s, the geodatabase was introduced to solve these data management challenges. Geodatabases now proliferate across the ArcGIS platform with a number of variations on the geodatabase model intended for individuals, workgroups, enterprise, and even large distributed organizations. At Grand Ronde, a number of tribal government offices have their data stored and centrally managed in a number of geodatabases on the GIS department's servers. The use of GIS varies across the tribal government offices. Some offices have only one or two occasional GIS users, while others, such as the NRD GIS Department, have four or five staff members using ArcGIS Desktop software on a daily basis. Across each tribal government office, GIS staff members are responsible for keeping their GIS information up to date. This is accomplished through a series of unique workflows where tribal government staff members edit data and conduct field observations to keep their information up to date. Despite the centralized management of data in the GIS department, the responsibility of keeping data current remains the primary function for each office based on its own mission within the tribal government.

One of the central geodatabases on the GIS server is dedicated to data that is commonly used by the natural resources department. There are several GIS layers stored in the geodatabase that are set up for versioned editing. Versioned editing allows GIS layers to be "checked out" from the master database for modification and updating. A versioned GIS layer might be taken into the field on a laptop and edited there based on field observations or modified locally while disconnected from the enterprise geodatabase. Each of the layers edited by NRD staff is assigned to a data steward, a reviewer who makes sure that edits made by the foresters and biologists to these layers are correct. This administrative step to authenticate the editing process assures that the incoming data into the GIS system is as current and accurate as possible and is a key part of the workflow. Once reviewed by an administrator, the versioned dataset is "checked in" to the master geodatabase.

Across an enterprise GIS framework, security is an important consideration. Not every tribal employee should have access to all of the GIS information in the system. In the case of Grand Ronde, access to the geodatabase is done through operating system authentication. The Grand Ronde IT Department manages several user-groups, which are used to control editing and viewing rights to the geodatabases. The Tribe also uses a geodatabase replication architecture to provide further security. In this way, multiple geodatabases are used in a hierarchy to allow concurrent multiuser editing while keeping the Tribe's master geodatabase secure. To accomplish this, a copy, or "child version" of the Tribe's master database, is created for the use of the NRD. Additional "children" databases are created for the use of the data stewards and subsequent child geodatabases for the NRD staff for editing. This model allows for additional security and control of local data as edits are rolled into the master database as approvals and reviews are completed. In the same manner, the versions edited by foresters and biologists are child versions of the data steward versions. This comprehensive system of versioning enables the Tribe to constantly edit all GIS data by the appropriate tribal employee.

To better understand how this GIS architecture serves the Tribe's information management needs, let's step through the GIS workflow that supports the planning of a standard timber sale. First, the NRD staff member opens a map document "Timbersale.mxd," which is stored on the file server. Foresters will usually create a file folder named after the timber sale to store temporary data, image files that are produced, and a subfolder storing data for field editing. ArcPad is a lightweight GIS application that runs on mobile devices, allowing GIS data to be updated in the field. Data checked out to ArcPad contains point

Chapter 1: Managing tribal resources and protecting the environment | 11

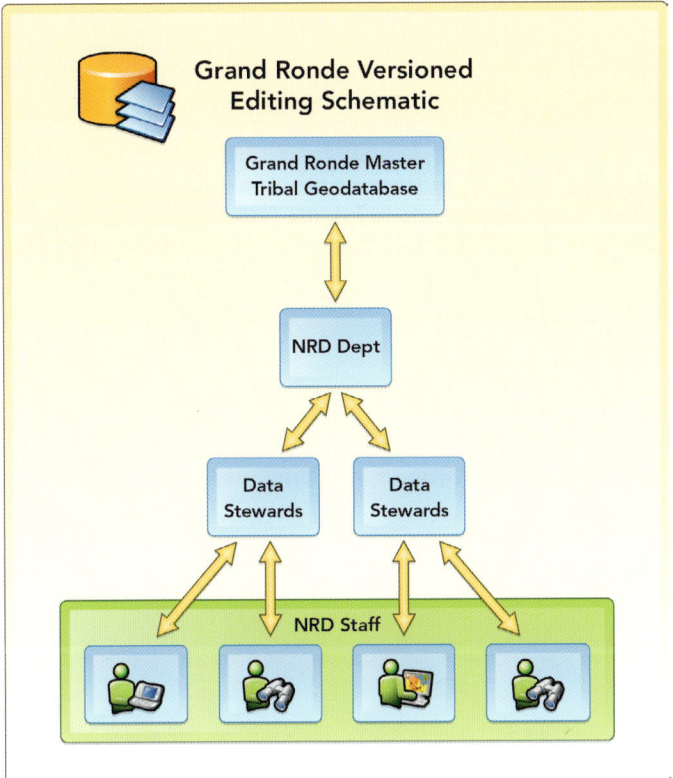

Figure 3. Grand Ronde NRD multiuser editing framework.

layers for roads, streams, and cut blocks, together with background data like aerial photos, contour lines, and the reservation boundary. The reason for not checking out versions of the original streams, roads, and cut boundaries is that due to a heavy canopy and steepness of terrain, it is not possible to collect accurate GPS data in all locations. The GPS availability, especially along streams, is spotty due to the even denser vegetation and the often restrictive topographic situation.

After the point data on proposed timber sales is acquired in the field, it is "checked back into" ArcGIS Desktop, using the ArcPad data manager extension. The GIS user then works with his or her version of the data to adjust streams, roads, and timber sale unit boundaries to the points surveyed with GPS in the field. The GIS user can then edit the data for better accuracy, relying on data layers such as high-resolution orthophotos and digital elevation models (DEMs) to amend the GPS data where the accuracy is questionable.

The data stewards who review the incoming data in this timber sale workflow are the timber and roads coordinator for the roads and timber sale boundaries; the biologist for the streams; and the silviculturist, or forester, for the stand boundaries. Edits done by multiple foresters in their versions of feature classes are verified against each other by the data stewards. The data stewards use the version changes tool in ArcGIS Desktop to compare the changes and decide if further field verification is necessary.

When the field edits are approved by the data stewards, they inform the GIS coordinator, who updates the master geodatabase. This versioned editing process of the central feature classes for the NRD improved the efficiency of the timber sale planning process. The data surveyed in the field is immediately available to all of the involved staff at NRD. Besides creating maps and updating existing GIS data, the NRD is using GIS to estimate the volume of board feet to be expected from a timber sale. After identifying the locations of all streams, roads, cutting block boundaries, and yarding roads, the forester can then use an array of geoprocessing tools, combined with the pre-sale timber cruise, to estimate the total volume removed from a cutting block. This is accomplished by removing non-forested areas from the cutting block so that an accurate estimate of board feet can be calculated. During this process, sensitive areas, such as elk meadows, can also be identified and protected from the commercial timber activities. The Grand Ronde regularly use geoprocessing tools, such as the buffer tool discussed earlier, and "cut polygon" and other tools, allowing the GIS representation to match the real world situation as closely as possible.

This use of GIS in support of the Grand Ronde NRD program makes it possible for the Tribe to manage the natural resources on the reservation in a comprehensive and holistic manner. The ecologic concept that everything is in some way connected in nature is showcased in the Confederated Tribes of the Grand Ronde's use of GIS to support the management of their natural resources for multiple use.

Courtesy of Navajo Division of Transportation

Moving ahead in transportation

As sovereign Nations, Tribes in the United States are responsible for all aspects of tribal government, including managing the transportation infrastructure for communities and tribal lands. The stories in this chapter demonstrate how several tribal governments have used GIS to meet the needs and challenges Tribes face with managing and maintaining their road systems. From managing road inventory, construction, and maintenance to permitting, planning, and routing, GIS is being used to help Tribes provide a safe and adequate roadway system, while saving money and contributing to the health, safety, and economic well-being of their communities.

A compelling reason to use GIS within any tribal transportation office is the Indian Reservation Roads (IRR) Program. In 1982, the Surface Transportation Assistance Act established the IRR. This program is jointly administered by the Federal Highway Administration's Office of Federal Lands Highway (FLH) and the Bureau of Indian Affairs (BIA). Indian reservation roads are public roads that provide access to and within Indian reservations, Indian trust land, restricted Indian land, and Alaska Native villages. The goal of the IRR Program is to address the transportation needs of Indian Tribes and Alaska Native villages by providing funds for planning, designing, construction, and maintenance of Indian reservation roads. The IRR Program funding for these activities is based on the accurate inventory of reservation roads. As demonstrated below, GIS can be a tremendously helpful tool in responding to the IRR Program. It plays a key role in the organization of road data, producing reports and maps, and in the planning of new and expanded roadways.

Leveraging GIS for transportation management provides a platform for decision support for other transportation-related programs. Some examples include planning more effective delivery for programs that use reservation roads, such as school buses, emergency vehicles, and delivery services. A GIS-based road network across a reservation enables the creation of response time maps for police and emergency departments and analysis of underserved areas. It also helps decision makers understand where new police and fire stations need to be placed as their community grows. School buses can be routed so they are using the shortest, most efficient routes to pick up and drop off schoolchildren, which saves them time, money, and fuel.

The stories for this chapter come from two different Tribes with large GIS implementations—the Department of Geospatial Information of the Chickasaw Nation and the Navajo Division of Transportation. Both of these stories demonstrate the important role GIS is playing in helping Tribes meet the challenges of tribal transportation.

Road network efficiencies using GIS at the Chickasaw Nation

John Ellis, Director, Department of Geospatial Information, the Chickasaw Nation
Ada, Oklahoma

The Road Network Project at the Chickasaw Nation began from a need to develop more accurate network response times and more efficient routes to save time, money, and, hopefully, lives. Before this project began, existing routes were accepted as the most efficient and best routes because they had been in use for a long time. With the support of GIS layers and analysis, it is now clear that the fastest or most efficient route isn't necessarily a straight line. Looking at a map, it is natural to visually calculate the route from point A to point C, assuming that going through point B is a given. Taking different elements into consideration, such as speed limit, construction areas, rush hour times, stop signs, and many other qualifying attributes, the quickest way from A to C may be through point D.

Another issue that exists within the realm of the transportation network is minimizing response time for emergency management services. The Chickasaw Nation has a large service area covering 7,443 square miles in south-central Oklahoma, which includes all or parts of thirteen counties. The roads network covers a staggering 14,598 miles of roadway. Within this area, the Chickasaw Government provides emergency services as well as transportation services. Using a response time map is an excellent tool for developing an optimal system of stations for a fire department, police department, or transportation office. The map can be used to locate the longest response times in order to determine placement of new stations. These response-time maps are used daily by the fire and police departments to ensure they are providing their services as economically and as effectively as possible to the Chickasaw Nation.

Developing a process for creating a better road network was the first step of the project. It was determined that the best way to store and manage this information would be in a geodatabase where a series of GIS layers would work together to represent the road network. The layers in this type of geodatabase are known as a feature dataset. Because there are hundreds of road networks available, it was necessary to first determine which existing road network was the most accurate, and then to use it as the base transportation feature dataset. Once this dataset was updated and corrected with new road data and additional attribute data, it served as the main dataset from which our other data layers and feature classes were derived. These include an IRR routes layer, a quick draw layer (for display on the web), and a display layer that is visually appealing at a higher scale.

Work on the transportation network in the feature dataset began by taking an existing road feature class, updating it with missing data, and adding new attributes needed for routing, such as drive time, pedestrian time, speed limits, height restrictions, from elevation, to elevation, hierarchy, and one-way restrictions. There were a total of 45,936 road segments to edit and update. New road features were added while analyzing new high-resolution aerial imagery. While it took a considerable amount of time to develop this transportation feature class, it was worth the investment.

After this long process was complete, the quick draw layer was created. The road network transportation feature class was copied, and many fields that were included for routing were deleted to allow the layer to process at a faster rate. After this feature class was finished, the information was also copied so that a display layer could be created from it. This new display feature class was made to allow for the creation of a more visually pleasing and less cluttered map at a larger scale. This feature class is also easier to label since many roads with two separate sides (such as interstates) show up as one road at the larger scale. The last feature class created was the IRR route layer. IRR routes were digitized from the roads feature class and depict all routes where the Chickasaw Nation assists with road maintenance.

As this project progressed, it was found that the same aerial data that was used at the start had to be used throughout in order to maintain consistency throughout the dataset. The aerials that were initially used to map one area did not have some of the new roads that were depicted in new imagery that was received six months after the start of the project. Unfortunately, the imagery didn't always line up correctly. This was handled by incorporating both imagery sources into the editing process. Roads are always changing and new subdivisions are always being built, so the newest dataset might not always include every existing element due to the time post-processing takes. Roads necessarily digitized from new imagery had to be rectified to the original imagery's correct spatial reference. Versioning was used for the first time at the Chickasaw Nation during this project. This allowed the unique instances of the same dataset to be edited and reconciled in workflows involving multiple editors.

To support this project, program personnel are organized in a versioned editing atmosphere with the GIS specialist at the center. The interns and the GIS specialist are the editors of the geospatial data, and the GIS specialist double-checks all geospatial data that is turned in to the GIS analyst. The GIS analyst is responsible for the Quality Assurance/Quality Control (QA/QC) process. Once all the data has gone through the QA/QC process, the data is then "posted" to the main (default) database to provide program-wide access.

The workflow for this project included using aerial imagery for visual analysis and digitizing features for the feature classes. Building the road network was another part of the main workflow, which included adding network attributes to the feature tables and building the network using the network analyst extension. This was done so that drive-time calculations and additional network-based analysis could

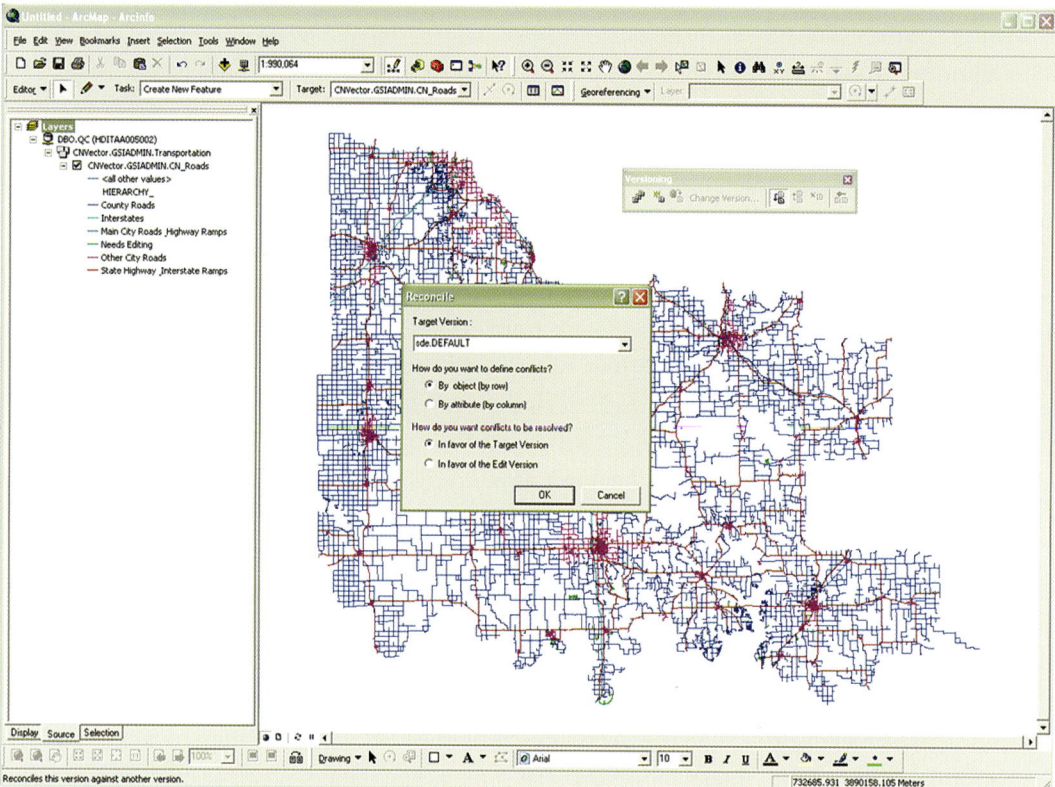

Figure 1. Versioned editing. Courtesy Chickasaw Nation

be performed. This part of the process was relatively easy compared to complications that came up related to the creation of datasets and data features.

This project took two years to implement, with interns working on the road network creation on a scale of approximately 1:1,500. Building the network took approximately two months. As a result, four critical layers have been created: a roads layer, a quick draw roads layer for the web, a display roads layer, and an IRR routes layer. There were three services resulting from the road network, including a bus routing service, a fire department response-time service, and a Lighthorse Police Department service. Other layers can also be derived from the road layer using the Network Analyst extension to ArcGIS Desktop.

Benefits from this transportation project have had an impact on many areas throughout the Chickasaw Nation. An accurate transportation network has helped emergency management crews respond more quickly. Transportation services for schoolchildren and others needing transportation are run more economically. Placement of offices and services to support a growing community is now done more effectively through careful and meaningful spatial analysis.

Chapter 2: Moving ahead in transportation | 17

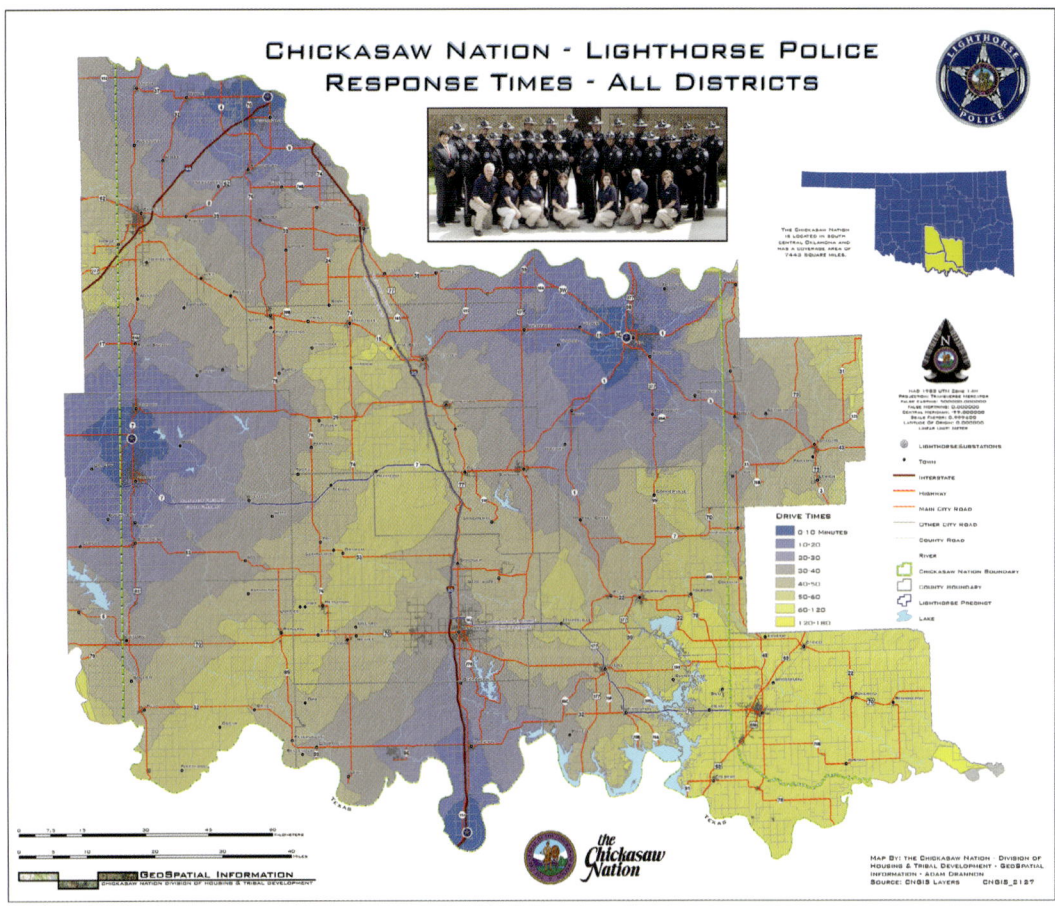

Figure 2. Lighthorse Police response-time map. Courtesy of the Chickasaw Nation, Division of Housing and Tribal Development, Geospatial Information, Adam Drannon

Using GIS to assess and manage tribal transportation infrastructure

Nick Hutton, Data Transfer Solutions, LLC; and Jonah Begay, IRR Inventory Coordinator, GIS Specialist, Navajo Division of Transportation
Window Rock, Arizona

Spanning approximately 27,000 square miles across three states, the Navajo Nation is the largest sovereign Nation in the contiguous United States. The Navajo Nation has a strong presence in the US government and often leads the way in tribal efforts to promote key areas such as economic development, health care, and education at the national level. Despite its prominence, the sheer size and remote nature of the Navajo Reservation presents unique challenges in managing its infrastructure and resources.

Consider, for instance, the road inventory that Tribes submit each year to the IRR Program. The IRR maintains the official inventory of reservation roads in the United States and is designed to allocate federal funding to tribal governments for transportation, planning, and road maintenance activities.

A component of the broader Integrated Transportation Information Management System (ITIMS) Program, the Bureau of Indian Affairs Division of Transportation (BIA-DOT) maintains the national reservation road inventory in a system called RIFDS—the Road Information Field Data System. Each year, as part of the IRR Program, Tribes are eligible to submit their road inventory data to one of the twelve BIA regional offices. There are approximately 560 nationally recognized Tribes that fall under the twelve BIA regions. The Navajo Nation submits its road inventory to the Bureau of Indian Affairs Navajo Regional Office (BIA-NRO) in Gallup, New Mexico.

The Navajo road inventory was far from comprehensive. In early 2006, its official RIFDS inventory contained approximately 9,800 miles of roads. Roughly, 6,000 miles were BIA roads, and the remaining 3,800 were primarily state and county roads, with very few tribal roads mixed in. Navajo transportation officials determined that the road inventory was substantially underperforming in two key areas.

- **Road mileage quantity:** The current inventory reflected only a small percentage of the reservation's tribal roads. It was widely believed that there were thousands of miles of tribal public roads that were eligible for the inventory, but were not yet included.
- **Data quality:** Of the 9,800 miles of roads in the 2006 inventory, only a portion generated funding in the RIFDS allocation formula. Some roads in the existing inventory were missing key pieces of information, which excluded them from funding. Misinterpretations of program regulations resulted in a lack of quality data, exacerbating the effect of the low mileage numbers.

To address these issues, in April of 2006 the Navajo Division of Transportation (DOT) launched a proactive and aggressive campaign that would expand its internal capacities, establish a systematic method for identifying eligible public tribal reservation roads, remove subjectivity from regulations, and build a system to improve both the quantity and quality of the road inventory data. With the support of the Navajo Nation Transportation and Community Development Committee (TCDC) and under the direction of the former Division of Transportation Director Tom Platero, Navajo DOT GIS Specialists Lemont Yazzie and Jonah Begay and consulting Project Manager Nick Hutton embarked on an innovative and challenging endeavor that would span more than four years.

The first step was to fortify the division's existing technology infrastructure. New enterprise-class servers were put into place, network bandwidth was expanded, and new data was collected. The Navajo DOT implemented a spatially enabled, multi-tiered, web-based information architecture that was part of an integrated hardware and software solution provided by the InLine Corporation (now IceWeb) and Esri. The IceWeb servers were preloaded with ArcGIS Server and MS-SQL Server, and were preconfigured to optimize system performance. This preconfigured system saved many hours of work by allowing the Navajo project team to focus primarily on the development of the core programs and data, instead of testing and tweaking the new system.

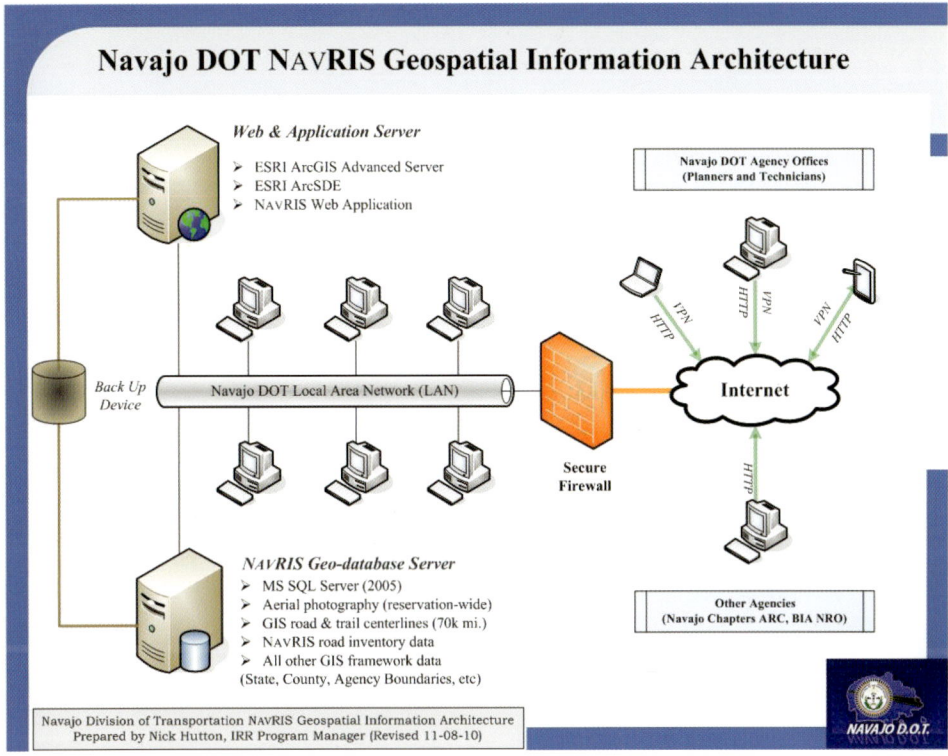

Figure 3. The NAVRIS geospatial architecture. Courtesy of Navajo Division of Transportation

The next step was to obtain and develop the required data. The project team was able to acquire brand new, reservation-wide aerial photography captured as part of a joint project between the Department of the Interior and the state of New Mexico. Once the imagery was loaded onto the new system, it was time to start digitizing road centerlines.

Along with a team of GIS technicians using Esri's ArcGIS technology, GIS consultant and Esri business partner Data Transfer Solution (DTS) began the digitization process. It wasn't until this time that the team realized the full extent of the project. After several months of heads-up digitizing, the team mapped more than 70,000 miles of roads and trails. While not all of the digitized centerlines were eligible for the official IRR inventory, the potential challenges associated with managing these roads were daunting to DOT officials. This realization underscored the notion that automation would be an absolute necessity in the development of the Navajo DOT road inventory system. While the GIS techs continued the digitization process, the programming staff at DTS and the Navajo DOT project team were busy developing the inventory management system.

The team concluded that the system must be secure, web-based, geospatially enabled, usable by staff members both with and without GIS expertise, and capable of mapping automation—specifically,

20 | Tribal GIS: Supporting Native American Decision Making

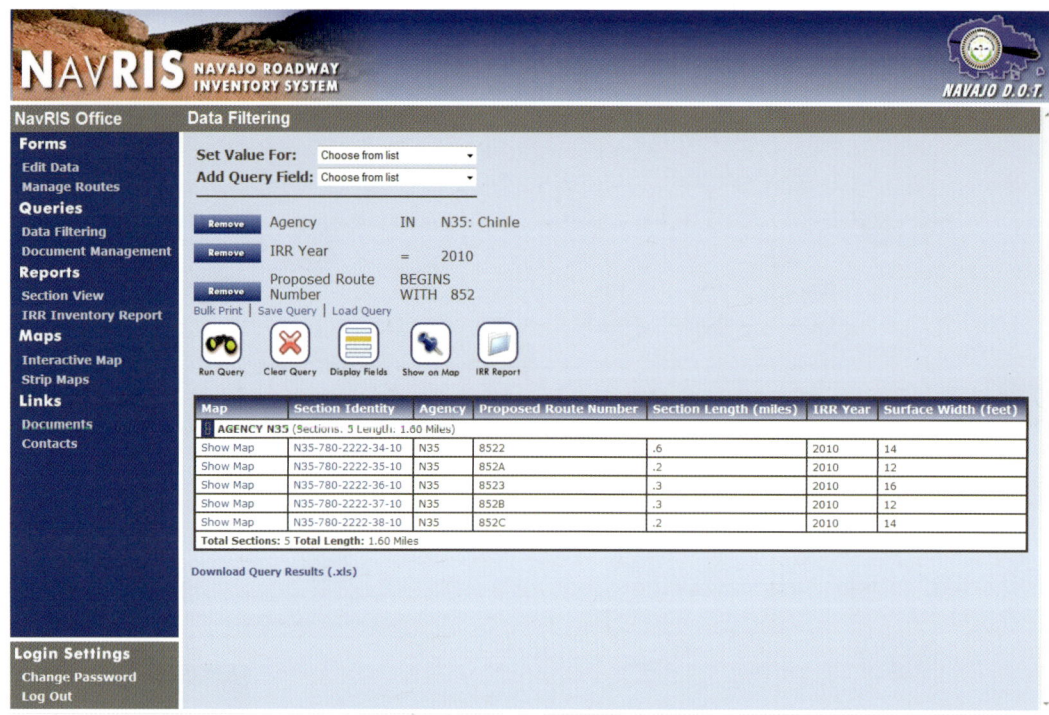

Figure 4. Bi-directional filtering of data between the map interface and the filtering page. Courtesy of Navajo Division of Transportation

Chapter 2: Moving ahead in transportation | 21

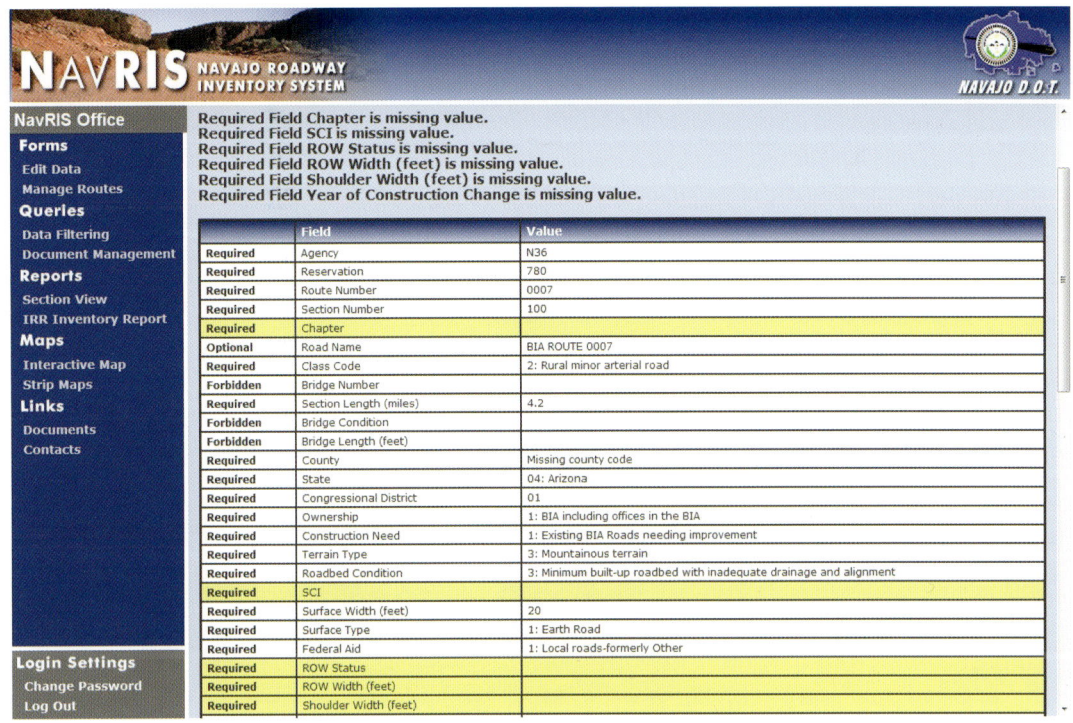

Figure 5. NAVRIS data validation. Courtesy of Navajo Division of Transportation

strip map automation. In addition, the team identified the need for a robust querying component that included bi-directional filtering between the map interface and the filtering page.

What emerged was a system the Navajo DOT calls NAVRIS—the Navajo Roadway Inventory System. In addition to web, GIS, and automation capabilities, NAVRIS incorporates a series of validation scripts to ensure the data is entered in accordance with program requirements.

One of the most challenging aspects of the project was establishing consistent interpretations of the IRR Program regulations between the BIA-NRO and the Navajo DOT staff. This took many months of research in collaboration with BIA-NRO Chief Engineer Harold Riley and his staff. To the credit of both agencies, considerable common ground was established and the findings were subsequently programmed into the core automation and validation logic of NAVRIS. As a result, the percentage of roads rejected by the BIA because of missing or incorrect data has declined dramatically.

As of the 2010 IRR submission cycle, the division staff has nearly doubled the number of miles in its inventory. It grew from 9,800 miles in 2006 to nearly 18,000 miles, including approximately 8,000 miles of tribal roads. The additional mileage and updates to the existing data increased the Navajo Nation's IRR funding by an average of 30 percent compared to its 2006 funding level. To date, the Navajo DOT has received a fifteen-fold return on its initial investment in the IRR project. This adjusted allocation will allow for critical transportation infrastructure improvements supporting access to education, employment, health care, and other services for the Nation's widespread residents.

Navajo Division of Transportation
Indian Reservation Roads (IRR) Program
"IRR Public Roads Identification Procedure"

NAVAJO D.O.T.

Appendix "A"

IRR Inventory Public Road Identification Checklist

The following checklist is a summarized "view" of the Public Road requirements as listed in the 25 CFR, Part170, Section 170.5. For a road to be considered public, it must meet at least two of the following criteria: (check all that apply)

Note: if only one item is checked yes, please provide further documentation/justification

		Yes	No
1	Does the road serve a public need? If yes, please define in remarks		
2	Does the road provide access to a cultural property or facility(s)		
	Examples (public facilities under 25 CFR Part 170.120 & 121):		
	Sacred Site		
	Medicinal Site		
	Gathering Site for Medicines		
	Gathering Site for Materials		
	Traditional Activity - Hunting		
	Traditional Activity - Fishing		
	Traditional Activity - Gathering		
	Traditional Activity - Other		
3	Is the road a Seasonal Transportation Route (non-recreational transportation route for access to seasonal locations) (under 25 CFR Part 170.123 & 124) If yes, please define in remarks		
4	Does the road serve as a Housing Access Road per 25 CFR Section 170.127 (3 or more homes)		
5	Does the road serve a public housing development (unless its privately owned)		
6	Is the road used as a school bus, public transport or mail route? If yes, please provide documentation		

Remarks:

Agency:	Route:	Section:	Date:	Inv Taker:

Navajo DOT Public Roads Guideline
Page 6 of 10

Figure 6. The Navajo DOT's public road identification checklist. Courtesy of Navajo Division of Transportation

In addition to the development of the NAVRIS system, the Navajo IRR project team also established a series of programmatic policies and standards to supplement the technology. Due to the rural nature of the reservation, determining the public eligibility of tribal roads has been a historically difficult process. In an effort to establish consistency in properly identifying a public tribal road, the Navajo DOT developed a public roads identification guideline that provides a checklist of characteristics that a road must contain before it can be considered public.

The division also established a methodology for determining the proper functional classification of roads, which was another area lacking clear guidelines between the bureau and the Tribe.

By creating NAVRIS and the supplemental policies and standards, the Navajo DOT has developed a systematic approach to maintaining its road inventory. Beyond supporting the immediate needs of the federal IRR program, NAVRIS serves as the foundation for a comprehensive infrastructure management system to support Navajo Division of Transportation activities.

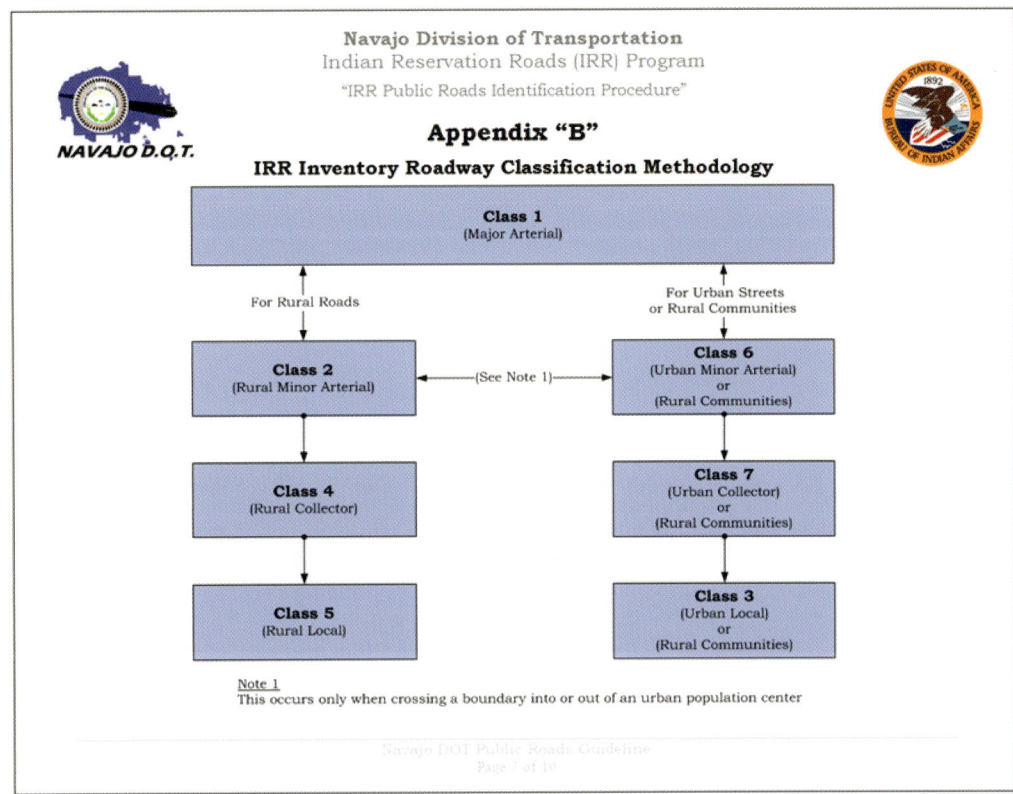

Figure 7. Public roads identification procedure. Courtesy of Navajo Division of Transportation

Today, the division continues to develop NAVRIS as part of its ongoing IT strategy. NAVRIS offers the first consistent, verified interpretations of IRR regulations and the ability to programmatically generate the required BIA deliverables. By taking the initiative to build a geospatial road inventory program that helps define and facilitate the IRR process, the Navajo Division of Transportation has become a stronger, more sophisticated tribal entity with more time and resources to support the development and maintenance of its expansive infrastructure.

Photo by Kirsten Grish

Preserving tribal culture and history

Understanding and preserving the past and linking it to the present are vital for the survival of a Tribe's culture today and for years to come. Knowledge of traditions roots new generations in the culture of their Elders, ensuring the continued practice and propagation of their traditional ways—even in the midst of modern competing interests. This chapter presents examples of how Tribes use GIS to help preserve and raise awareness about their cultural resources, traditional place names, cultural geographic boundaries, and cultural identity. The use of maps to identify sacred areas or cultural sites is an obvious use of GIS, but these contributors have taken the next step, using GIS to manage and analyze sensitive historical data so strategic decisions can be made to protect cultural sites, cultural traditions, and the sustainability of their Nations. Contributors to this chapter include GIS professionals from the Seminole Tribe of Florida, the Round River Conservation Studies Organization, the Confederated Salish and Kootenai Tribes, and the Chickasaw Nation. Contributors also include a geography professor and a master's student at San Francisco State University, and the Headman of the Winnemem Wintu Tribe.

Seminole geography: Using GIS as a tool for tribal historic preservation

Juan J. Cancel and Paul N. Backhouse
Seminole Tribe of Florida, Tribal Historic Preservation Office
Big Cypress Reservation, Southern Florida

In recent years, the management of North American indigenous cultural heritage has been substantially strengthened by the incorporation of geographic technologies within the fabric of tribal government infrastructure. The examples presented in this paper are drawn from the practical implementation of geographic information science (GISci) within the Seminole Tribe of Florida Tribal Historic Preservation Office (THPO). Specifically, the office sought to implement a data management system that describes the geographic cultural boundaries of Native Peoples (both through time and across space); presents different perspectives on past events using historical GIS; and integrates archaeological databases, oral histories, and cultural traditions with scientific tools (for example, GPS and ground-penetrating radar, or GPR). This approach brought about a practical, sustainable, and easy-to-operate system for the scientific management of the Tribe's heritage resources.

The Seminole Tribe of Florida (STOF) is a federally recognized Native American Tribe that has a long and complex history in the Southeastern United States (Covington 1993; Fairbanks 1974; Sturtevant 1987; Weisman 1999; Wright and Leitch 1986). Seminole People have a rich cultural heritage and are proud of their hard-earned "unconquered" status, despite more than four centuries of colonial encroachment into their territory (Weisman 1999). Historical sources, oral traditions, genetic data, and the archaeological record demonstrate that the Tribe has often shifted its geographic focus during both prehistoric and historic times (Williams and Shapiro 1990). From the late 1980s onward, the Tribe has publicly shown an active interest in the preservation of its historical and cultural resources, although informally this interest is undoubtedly much older.

GISci is particularly appropriate to the function of preserving tribal history. Anthropologists have long drawn from geographic perspectives to examine complex settlement systems that result from human subsistence (for example, Kantner 2008; Knapp and Ashmore 1999). The application of this concept to the history of the Seminole People highlights the dynamic nature of their lifeways, which have rapidly and often necessarily adapted to changing environmental and social conditions. Modern reservations are dynamic cultural landscapes, which exhibit a vibrant mix of cultural elements that range from modern construction and large-scale public works projects, to tribal members' home sites and areas of traditional plant gathering. Each of these landscape settings can hold unique symbolic meaning to individuals, clans, or the Tribe as a whole (*sensu* or "in the sense of," Basso 1996).

The STOF Tribal Historic Preservation Office (STOF-THPO) is the federally recognized tribal agency responsible for the protection of the Tribe's cultural heritage. Under the direction of the THPO officer, one of the responsibilities of a THPO is to review federal actions under Section 106 of the National Historic Preservation Act (NHPA). Geographically, the consultation process extends to all lands considered by the Tribe to be ancestral, aboriginal, or ceded (NHPA, 1966, Section 101; and 36 Code of Federal

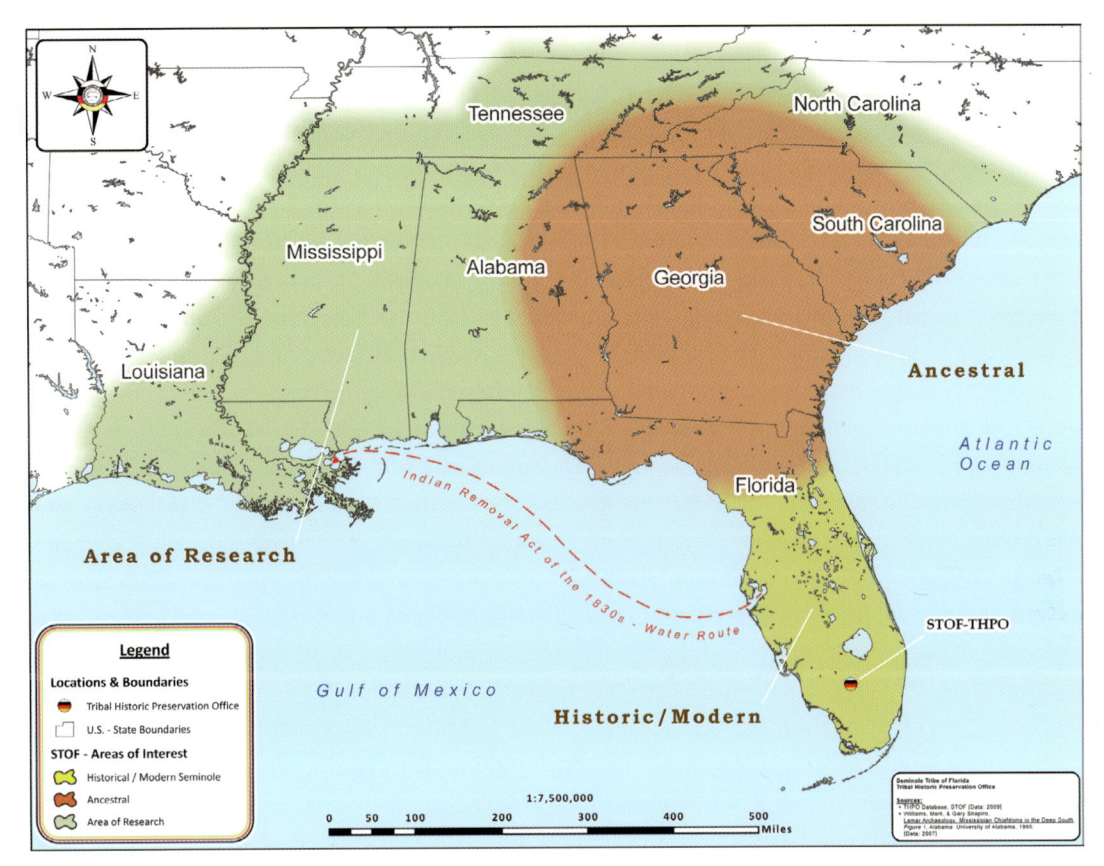

Chapter 3: Preserving tribal culture and history | 27

Figure 1. Areas of ancestral interest for the Seminole Tribe of Florida. Courtesy of Seminole Tribe of Florida, Tribal Historic Preservation Office

Regulations [CFR], Section 800.2). In addition to this legal responsibility, Native American Tribes have a vested interest in the discovery, maintenance, and dissemination of records pertaining to their own history. The Seminole Tribe of Florida is no exception.

From the outset, serious consideration was devoted to the development of GISci as a core component of the STOF-THPO program. Archaeology and history, after all, are both disciplines with methodologies that do not necessarily require the use of GIS. Nevertheless, the STOF-THPO has adopted a deeply geographic perspective because the ability to use visual and analytical tools that facilitate the interrogation of spatial datasets is central to the development of models of cultural adaptation (Conolly and Lake 2006; Kantner 2008; Wandsnider 1998). For example, initial examination of settlement data gathered from archaeological sites on the Brighton Reservation suggests that patterning can be identified in the relationship of habitation areas to two ancient watercourses that run through the reservation.

The result of the foregoing example might be archaeologically predictable. However, the interface between the Seminole People and their cultural environment requires a quite different epistemological perspective than is typically employed by archaeologists. It is felt that a willingness to develop and adapt a program of GISci at a pace that is in tune with the Tribe's cultural practices and beliefs is very important.

Subsurface investigations and artifact collection are therefore not always the best courses of action. For archaeologists, this appears to be somewhat of a paradox as they are trained to disturb the earth in order to record, observe, and interpret. The role of the tribal archaeologists is fundamentally different; the job of the THPO is to preserve cultural remains *without* significantly impacting sites. As a result, GIS has been crucial for implementing a "hands-off" approach to areas relating to or potentially relating to Seminole ancestors.

The STOF-THPO is divided into six separate operational sections. Thanks in large part to a standing licensing agreement with the Bureau of Indian Affairs, a full suite of GIS (ArcGIS from Esri) products is available to all of these sections. The realization that a heavy demand for geographic data was inherent in the office structure and that this information would often need to be required to be shared between sections underscored the need to organize GIS data in a logical fashion. The solution has been the development of a local GIS server. The server houses geodatabases that are accessible to office staff on a permissions basis. Security of culturally sensitive data is a major concern, and the THPO server runs parallel to the wider enterprise level GIS architecture that has been established for tribal government. Examples of the geospatial layers contained within the database include the spatial extents of the Seminole Reservations, the location of archaeological sites of importance to the Seminoles on and off reservation, areas of high research value, and georeferenced historical maps. The concept behind the server structure is that GIS is inclusive within the daily routines of all office personnel, instead of being the exclusive domain of GIS technicians and administrators. In order to achieve this goal, a diverse series of task types and operator skill levels needed to be accommodated, and outputs must be structured so that they are of use to nonspecialists in other areas of the office.

Tribal archaeology is perhaps the most operationally critical section. The archaeological research method employed by the Tribal Archaeology Section is not typical of the wider discipline, in that fieldwork is for the most part noninvasive; in other words, it uses techniques that do not disturb the ground. From a technology standpoint, the Tribal Archaeology Section GIS staff typically uses GPS and remote sensing to achieve the goals of minimum disturbance to cultural resources. GIS is used to create and record feature classes such as shovel tests, site boundaries, and proposed land uses. In the field, mobile GIS data acquisition is typically used to accurately map site boundaries. Using this methodology, the exact area of interest can be tracked very precisely, which can be further analyzed and documented back in the office.

Recording and preservation of tribal cultural properties (TCPs) (see King 1999, 2003; and Parker and King 1998 for a discussion of this resource type) is of primary importance to the mission of the STOF-THPO because they represent areas that are vital to the cultural identity of the Seminole People. TCPs are inherently geographic; they exist in particular places, but they are also dynamic. They may move with the rhythm of seasonal variations or become redundant within short intervals of time. Data capture is again based around a mobile GIS solution. The methodology, however, is quite different to that being applied for archaeological fieldwork. TCPs can be identified by many different mechanisms that do not necessarily begin with fieldwork. Examples of identification procedures include formal oral histories, chance encounters with other tribal members, and activities associated with the daily life of tribal members on the reservation. Highly portable and rugged devices such as Trimble's Juno handheld system are ideal for on-the-fly data acquisition and recording. These devices are used by the cultural staff of

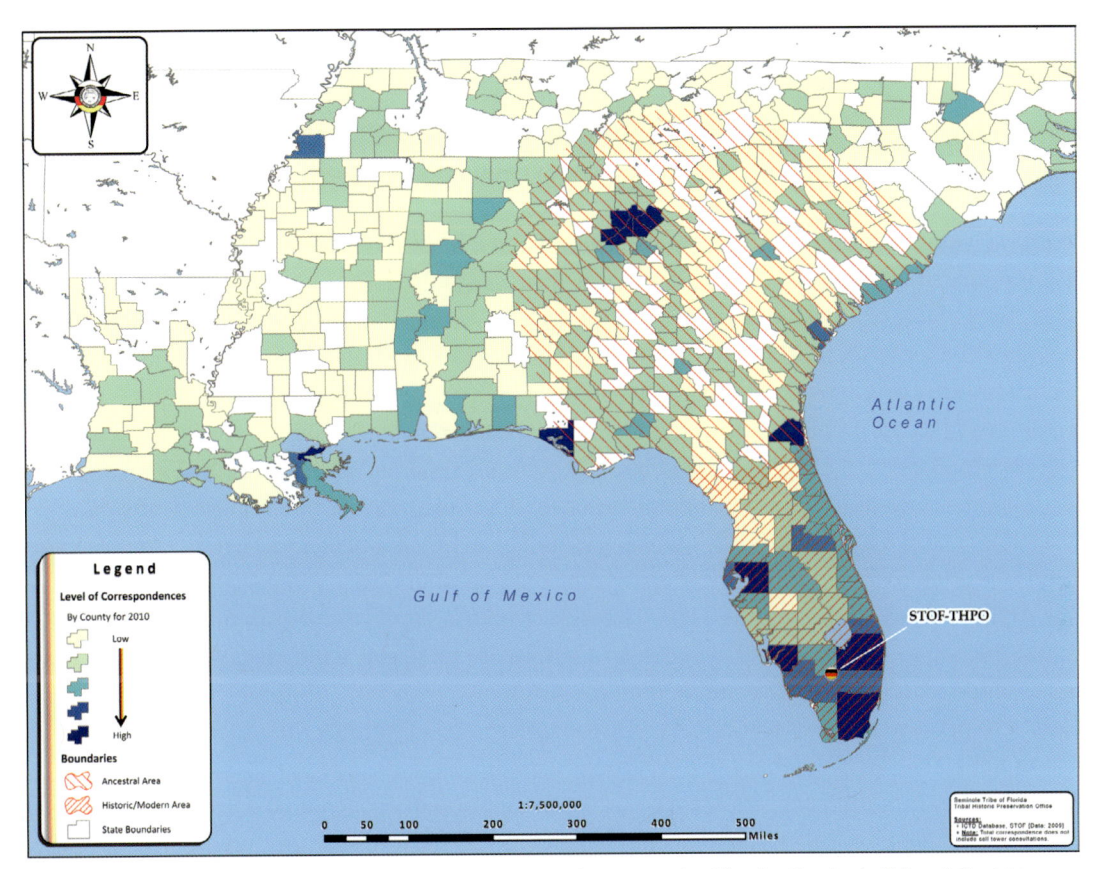

Figure 2. Analysis of the frequency of federal correspondence received by the Seminole Tribe of Florida THPO (data, 2010). Courtesy of Seminole Tribe of Florida, Tribal Historic Preservation Office

the THPO and are downloaded at regular intervals to update the cultural geodatabases layers stored on the THPO server. The records for TCPs, including GIS layers, are restricted with secure access privileges for particular personnel. Cultural GIS layers are critical in the broader tribal planning process. A good example of the success of this program has been in coordinating efforts to treat invasive plant communities with herbicides while avoiding traditional plant gathering areas in use by tribal members.

The Compliance Review Section oversees consultation with federal, state, and tribal agencies to see that the legal responsibilities of federal statutes, especially NHPA (1966), the Native American Graves Protection and Repatriation Act (NAGPRA), and the Archaeological Resources Protection Act (ARPA), are being adhered to. GISci is critical to this section of the office because each project submitted for review must be compared to areas of land considered by the Tribe to be of ancestral interest (see figure 1). Project data under review by compliance staff can be compared, on the fly, with the state archaeological site file information in order to address areas of concern, which may impact cultural resources. Additionally, an incoming correspondence database contains GIS elements, which provide feedback on the geographical locations of federal agencies sending information to the STOF-THPO (see figure 2). In this way, agencies can be quickly identified, which may not be aware that they should be sending projects for review to the STOF-THPO.

The Collections Section attempts to connect the historical record of Seminole and ancestral occupation in the Southeast with the archaeological record and oral traditions within the Tribe. A crucial example of the use of GIS within this section is the creation of a Seminole Site File (SSF). The SSF is a list of all the cultural resource locations in the Southeastern United States that are relevant to the Seminoles or their ancestors. The SSF takes into account both its own STOF-THPO records and also six applicable state site files and historical records. This will be the first effort to draw back together the disparate archaeological evidence for the Seminole People within the context of the ancestral settlement systems of the Southeastern portion of North America. The scale of this task is immense and requires the study of approximately 130,000 archaeological sites spread over a six-state area of the southeastern United States (Backhouse et al. 2007). Not only are these records voluminous, but they have also been recorded differently and housed at state level repositories (Anderson and Horak 1995). This is important work, though, because it is critical that these region-spanning records can be related in a consistent way, over several eventful centuries, to form a holistic understanding of the past of Native groups such as the Seminole.

The Archaeometry Section of the THPO is tasked with the deployment and management of GIS. In addition, this section undertakes fieldwork and laboratory analyses, which require scientific theories, methods, and techniques to solve archaeological problems. A key role, therefore, of the Archaeometry Section is to manage GIS projects initiated by the other sections within the office, and it was for this reason that the THPO GIS database server structure model was developed.

A good example of this section's activities is the integration of geophysical data with on-reservation project GIS. The geophysical techniques employed by the THPO are typically subsurface and are used to measure phenomena that are not visible to the instrument (Johnson 2006). Background research as to the types of archaeological signatures associated with Seminole subsistence strategies and the burial environments in south central Florida strongly suggested the acquisition of GPR as the primary remote-sensing device.

Combining the GPR data with GPS and integrating the resultant plots within a GIS can be problematic because all three systems use different software coding and architecture. Nevertheless, both the GPR and GPS are inherently geographically based systems operating within a common spatial environment. By using a serial recording unit, a high-end, multi-channel Global Navigation Satellite System (GNSS) GPS receiver can be connected with the GPR, so that extremely accurate real world coordinates for any geophysical survey area can be obtained. This data, in combination with the databases and geodatabases maintained by the office, allows users to query and manage the raster and vector layers generated by the field hardware. Statistics and spatial analysis can subsequently be developed to examine if culturally relevant spatial patterning is observable in the dataset.

Historic Architecture is the sixth section of the STOF-THPO. This section deals primarily with the built environment on the reservations. Practically, this translates to researching and documenting buildings and structures of key importance to Seminole heritage and can include both on and off reservation reviews. Integration with the GIS server is again important to this section, as the building resource locations are plotted and maintained within a context of the human and physical geography that comprises the landscape of the modern reservations. The Historic Architecture Section is conducting a comprehensive survey of traditional chickees (thatched wooden structures) across all the Tribe's reservation lands. The survey uses a customized digital recording form uploaded onto a Trimble GeoXT handheld GPS device to record various aspects of these structures, including the style, size, and linked photographic record. The cultural use of chickees has adapted over the years, and the results of this survey will be the first to examine the myriad new styles and uses for which these enigmatic structures are now being built. For this project, GIS helps to organize the collected information geographically, so that tribal members and staff can have an accurate record of the role of chickees within Seminole culture in the early part of the twenty-first century.

It is hoped that the preceding brief overview of the STOF-THPO has highlighted some of the aspects in which geography, in particular GIS, can be integrated into the operations of a THPO. Plans for the future are to build upon the solid geographic base that has been developed. Completion of the Seminole site file will give tribal members access to and control over the cultural resource information that has, up to this point, been curated in federal facilities and in states hundreds of miles away from the modern location of the Tribe. The Archaeometry Section intends to teach classes on GISci to build a skill set that facilitates the development of a legacy spatiotemporal geodatabase for the Seminole Tribe that integrates all aspects of the available environmental, cultural, and historical datasets. Of course, training tribal members to use GIS is the first step towards what hopefully will be a contribution to Seminole geography carried out by tribal members.

Acknowledgments

The authors would like to thank David Gadsden and Anne Taylor (Esri), Willard Steele (STOF Tribal Historic Preservation Officer), and Anne McCudden (Ah-Tah-Thi-Ki Museum) for their useful suggestions and discussions, which have greatly improved the quality of this text. Any errors, however, are the responsibility of the authors. This text represents part of the ongoing research undertaken by the Seminole Tribe of Florida THPO.

References

Anderson, David G., and Virginia Horak, eds. 1995. "Archaeological Site File Management: A Southeastern Perspective." *Readings in Archeological Resource Protection Series 3*. Atlanta: Interagency Archeological Services Division, National Park Service.

Backhouse, Paul N., Marion F. Smith, Jr., and Juan J. Cancel. 2007. "Across State Lines: Using State Databases to Explore Seminole Ethnogenesis in the Southeast." Paper presented at the 64th Annual Southeastern Archaeological Conference, Knoxville.

Basso, Keith H. 1996. *Wisdom Sits in Places: Landscape and Languages Among the Western Apache.* Albuquerque: University of New Mexico Press.

Conolly, James, and Mark Lake. 2006. *Geographical Information Systems in Archaeology*. Cambridge: Cambridge University Press.

Covington, James W. 1993. *The Seminoles of Florida*. Gainsville: University Press of Florida.

Fairbanks, Charles H. 1974. *Ethnohistorical Report on the Florida Indians.* New York: Garland.

Johnson, Jay K., ed. 2006. *Remote Sensing in Archaeology: An Explicitly North American Perspective.* Tuscaloosa: University of Alabama Press.

Kantner, John. 2008. "The Archaeology of Regions: From Discrete Analytical Toolkit to Ubiquitous Spatial Perspective." *Journal of Archaeological Research* 16: 37–81.

King, Thomas F., comp. 1999 *Section 106: An Introduction.* Alexandria: National Preservation Institute.

———. 2003. *Places that Count: Traditional Cultural Properties.* Walnut Creek, California: AltaMira Press.

Knapp, Bernard A., and Wendy Ashmore, eds. 1999. "Archaeological Landscapes: Constructed, Conceptualized, Ideational." In *Archaeologies of Landscape: Contemporary Perspectives* Oxford: Blackwell Publishing, 1999, 1–32.

Parker, Patricia L., and Thomas King. 1998. "Guidelines for Evaluating and Documenting Traditional Cultural Properties." *National Register Bulletin 38.* National Park Service, US Department of the Interior.

Sturtevant, William C., ed. 1987. *A Seminole Sourcebook* London: Taylor and Francis.

Wandsnider, LuAnn. 1998. "Regional Scale Processes and Archaeological Landscape Units." In *Unit Issues in Archaeology*, edited by Ann F. Ramenofsky and Anastasia Steffen, 87–102. Salt Lake City: University of Utah Press.

Weisman, Brent R. 1999. *Unconquered People: Florida's Seminole and Miccosukee Indians.* Gainsville: University Press of Florida.

Williams, Mark, and Gary Shapiro. 1990. *Lamar Archaeology: Mississippian Chiefdoms in the Deep South.* Tuscaloosa: The University of Alabama Press.

Wright, Jr., and J. Leitch. 1986. *Creeks and Seminoles: The Destruction and Regeneration of the Muscogulge People.* Lincoln: University of Nebraska Press.

Taku River Tlingit First Nation and Round River conservation studies: Decision Support Tool

Richard Tingey, Senior GIS Analyst
Round River Conservation Studies, Salt Lake City, Utah

The Taku River Tlingit First Nation is located in Atlin, British Columbia (BC), Canada. Atlin is the northernmost town in BC, located at 59° 35' north latitude, 133° 43' west longitude, northeast of Juneau, Alaska, and southeast of Whitehorse, Yukon. Round River Conservation Studies is located in Salt Lake City, Utah, and is currently engaged in First Nations' land planning in BC.

The lands of the Taku River Tlingit First Nation (TRTFN) encompass the watersheds of the Taku and Whiting Rivers, as well as the headwaters of the Yukon River. The richness of this land and its rivers provides the very foundation of the Tlingit's *khustìyxh*, "or way of life." The Tlingit take their name, the *Takhu Quan*, from the Taku River itself.

Over 95 percent of the 10 million-acre territory of TRTFN in northwestern BC is wilderness. The prevailing salmon producer of southeast Alaska and the largest intact wilderness river system on the Pacific

Coast of North America, the 4.14-million-acre watershed of the Taku River dominates this territory as it flows from the interior boreal forests of BC to the coastal temperate forests of Alaska.

In the early 1990s, the government of BC initiated a province-wide initiative to designate protected areas and to create a series of strategic land-use plans to provide certainty for all resource users with regard to resource access, use, and conservation. The resulting Land and Resource Management Planning (LRMP) process brought together a range of stakeholder interests to negotiate agreements, and became the primary focus for conservation efforts across BC for much of that decade. There was limited participation by First Nations in the LRMP processes, or in the other landscape-level planning processes that followed, due to concerns that multi-stakeholder planning was ill-equipped to deal with aboriginal rights and title issues and that the process did not afford First Nations appropriate recognition as sovereign Nations. Nonetheless, by the early 2000s, provincial land-use planning had been completed for the vast majority of areas in BC, and had set in place a suite of land-use designations with varying levels of both protection and opportunities for industrial access and development. One of the very few remaining large and ecologically rich regions for which a strategic land-use plan had *not* been completed, however, was the Atlin-Taku, the territory of the Taku River Tlingit First Nation.

Recently, the BC government embraced a systematic, science-based planning approach as the foundation for its land and marine management. For example, in the central and north coast regions of BC, conflict between the timber extraction industry and environmental concerns stalled land-use decisions, so the BC government, timber industries, and environmental organizations agreed to jointly cooperate and support a regional-scale, science-based conservation area design developed by a coalition of independent scientists, called the Coast Information Team (www.citbc.org 2004). During that same time, in northeastern British Columbia, the BC government completed a systematic conservation assessment for the Muskwa-Kechika Management Area to ensure that management actions supported regional and sub-regional conservation goals for biodiversity maintenance and ecological integrity (Heinemeyer, Tingey et al. 2004).

One of the most important advancements in BC land planning in the past twenty-four months was the establishment of formalized land-use planning negotiations between the Taku River Tlingit First Nation Government and the government of British Columbia (government-to-government G2G), to support the Atlin-Taku Land Use Plan (AT LUP). The initiation of these talks is the product of a number of circumstances, not the least being the TRTFN's own determination to assert their rights and title through the Canadian courts. This, combined with the TRTFN being equipped with their own robust Conservation Areas Design and Land Plan, has generated a unique opportunity to create a lasting conservation outcome for their territory, through participation in the AT LUP. Over the next two years the TRTFN will seek resolution with the province to designate protected areas, establish authority to make decisions for resource management, and promote a sustainable economic future for their territory. Few other First Nations in Canada have been able to prepare themselves for engaging in a joint planning process by bringing together the range and quality of technical products that have been compiled by the Taku River Tlingit. Currently, Round River Conservation Studies is intensively engaged in providing technical support to the TRTFN as they negotiate an ambitious land-use agreement with the BC government for this 7.5-million-acre piece of their 9.8-million-acre traditional territory.

Beginning in 1998, the TRTFN proactively partnered with Round River Conservation Studies to begin laying the groundwork for a comprehensive land planning process for the territory—a process that not only included First Nation People, but invited agencies, governments, and the public to participate. The work included several years of information gathering, mapping, research, and interviews with Band members and the non-Tlingit people living in the community. This information gathering enabled the completion of the first defensible and systematic assessment of the ecological, social, and cultural values within traditional territory, the "TRTFN Conservation Area Design" (Heinemeyer, Lind, Tingey 2003).

In the years following the completion of the original TRTFN Conservation Area Design (2003), new and revised GIS data products have been made available by BC's Integrated Land Management Bureau (ILMB). The desire to incorporate new, improved spatial data into the current planning process prompted the TRTFN and BC governments to engage in the development of an all new and *jointly developed* assessment of the biodiversity and cultural values of the region.

To facilitate the regional assessment, and the creation and quantitative analysis of draft land-use scenarios developed by the Technical Working Group, a GIS-based Decision Support Tool (DST) was developed. This allows for the integration of data and analytical methods consistent with contemporary Western science with information related to the traditional use of the territory by the Tlingit. To date, the Atlin-Taku DST has made the decision-making process more transparent and increased the opportunities for both parties to collaboratively identify land-use solutions that meet their constituents' interests. The DST allows rapid review and analysis of a wide range of cultural, ecological, and socioeconomic values across the land base. This allows stakeholders to quantify the jointly developed land-use designation scenarios (potential land-use zone boundary configurations), to see if they meet mutual objectives for resource representation and spatial configuration across the landscape, thus ensuring that the *shared* objectives of cultural, ecological, and economic sustainability are met.

Building on the foundation of the TRTFN Conservation Area Design and decades of Tlingit cultural research, the DST is used to highlight the highest value ecological and cultural landscapes in the territory and generate maps that have played key roles in the technical G2G process, as well as the TRTFN and local community and stakeholder engagement process. As an example, a key product of the tool has been the High Value Landscapes for Tlingit Cultural and Ecological Values Map (CEV), which identifies the most valued areas as defined by their combined cultural and ecological values.

Using the DST tool, and following an intensive community consultation process, the TRTFN have now publicly released their vision for a network of protected lands across the planning area. Called Tlatsini (*Klat-see-nee*), which means "places that make us strong," the configuration of protected lands is intended to protect *khustiyxh*—the Tlingit land-based way of life—and to provide a reasonable balance of opportunities for a range of land uses and development (see figure 3). This map identifies the most ecologically and culturally important areas in the territory and highlights some 55 percent of the planning area.

The two-year process to develop a formal joint land use plan for the Atlin-Taku region is at a critical stage as the TRTFN are currently negotiating the Tlatsini areas and other land use zones and management direction. Completing a substantive negotiation of a land use agreement is anticipated by the end of 2010.

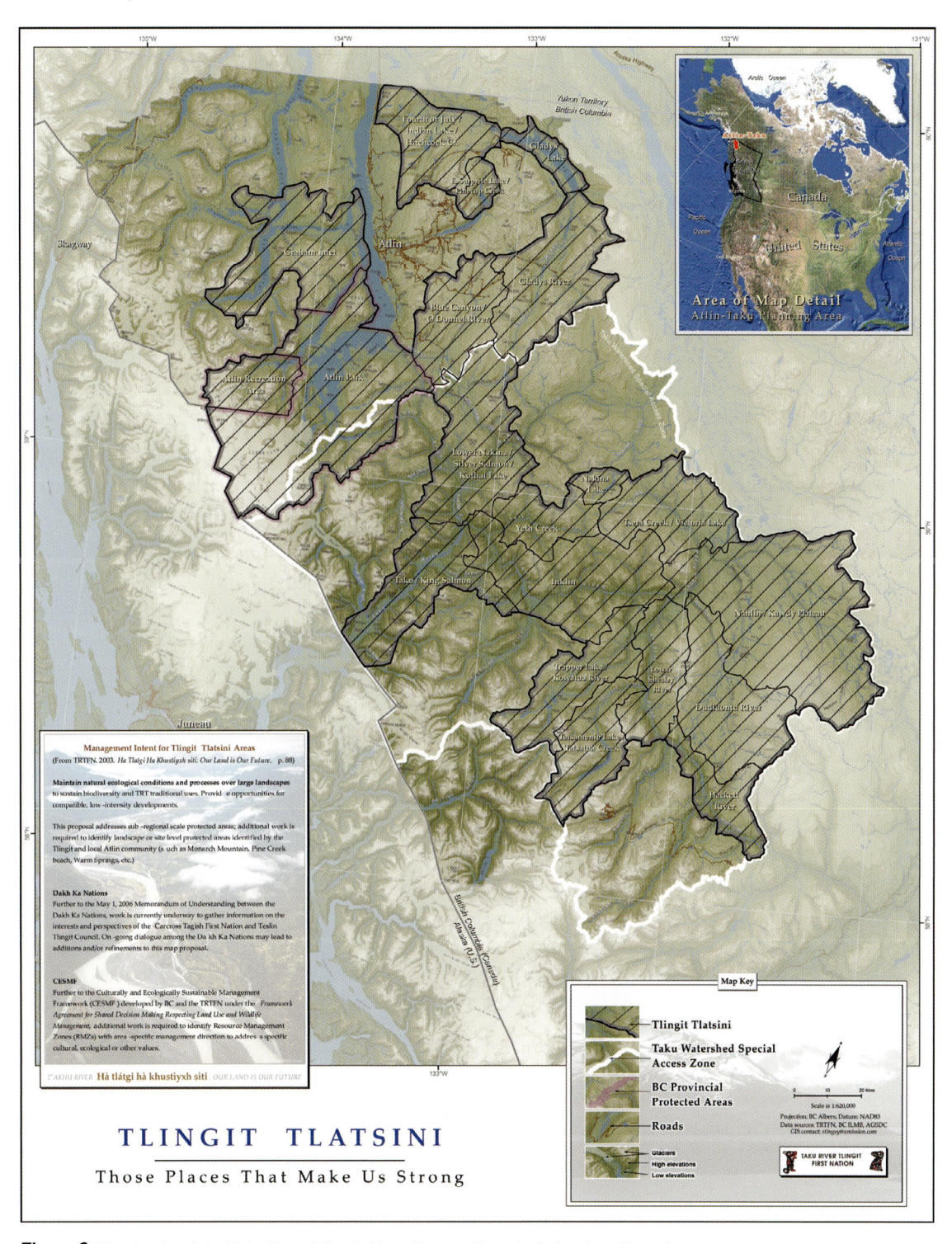

Chapter 3: Preserving tribal culture and history | 35

Figure 3. The lands of the Taku River Tlingit First Nation, British Columbia, Canada. Courtesy of Rick Tingey, Round River Conservation Studies

Equally important as the development of spatial land use zones, a parallel agreement is under negotiation that will establish the rules around the shared decision-making processes that the two levels of government will follow to implement this land use agreement and collaborate into the future. Here as well, the DST is proving to be a valuable asset in exposing the values of any given place when a land use decision is being made. The government-to-government decision-making agreement will be a cornerstone for the Tlingits, ensuring that they are equal partners with British Columbia in stewarding the rich lands and resources of their territory, and promoting sustainable development that benefits future generations of Tlingits and the local Atlin community.

On July 19, 2011, the TRTFN signed historic agreements with the government of British Columbia establishing land protection measures and shared management responsibility for ancestral lands. The *Wóoshtin Wudidaa* (Flowing Together) Land Use Plan protects more than seven million acres from commercial logging and designates over two million acres as First Nation Conservancy Parks. In addition, the Taku River Tlingit and provincial government have agreed to a joint governing process, *Wóoshtin Yan Too.aat* (Walking Together), to guide future resource-related decisions.

References

Coast Information Team. 2004. http://www.citbc.org.

Heinemeyer, K., T. Lind, and R. Tingey. 2003. *A Conservation Area Design for the Territory of the Taku River Tlingit First Nation: Preliminary Analyses and Results.* A report prepared for the Taku River Tlingit First Nation. Round River Conservation Studies, Salt Lake City, Utah, 98.

Heinemeyer, K., R. Tingey, K. Ciruna, T. Lind, J. Pollock, B. Butterfield, J. Griggs, P. Iachetti, C. Bode, T. Olenicki, E. Parkinson, C. Rumsey, and D. Sizemore. 2004. *Conservation Area Design for the Muskwa-Kechika Management Area.* Prepared for the BC Ministry of Sustainable Resource Management.

Confederated Salish and Kootenai Tribes: The Flathead Indian Reservation land status map with tribal place names

William Fisher, GIS Analyst, Confederated Salish and Kootenai Tribes NRD/GIS Department, Pablo, Montana

The home of the Confederated Salish and Kootenai Tribes (CSKT) is the Flathead Indian Reservation, 1.317 million acres situated in northwest Montana. This Nation comprises the Bitterroot Salish, the Upper Pend d'Oreille, and the Kootenai Tribes. The ancestors of the present-day Tribes inhabited the territory now known as western and central Montana, parts of northern Idaho, southern British Columbia, and northwest Wyoming. This aboriginal territory exceeded 20 million acres at the time of the 1855 Hellgate Treaty. The treaty ceded much of this territory and reserved the area now known as the Flathead Reservation for a homeland. Through unilateral legislation of the US government, much of the Flathead

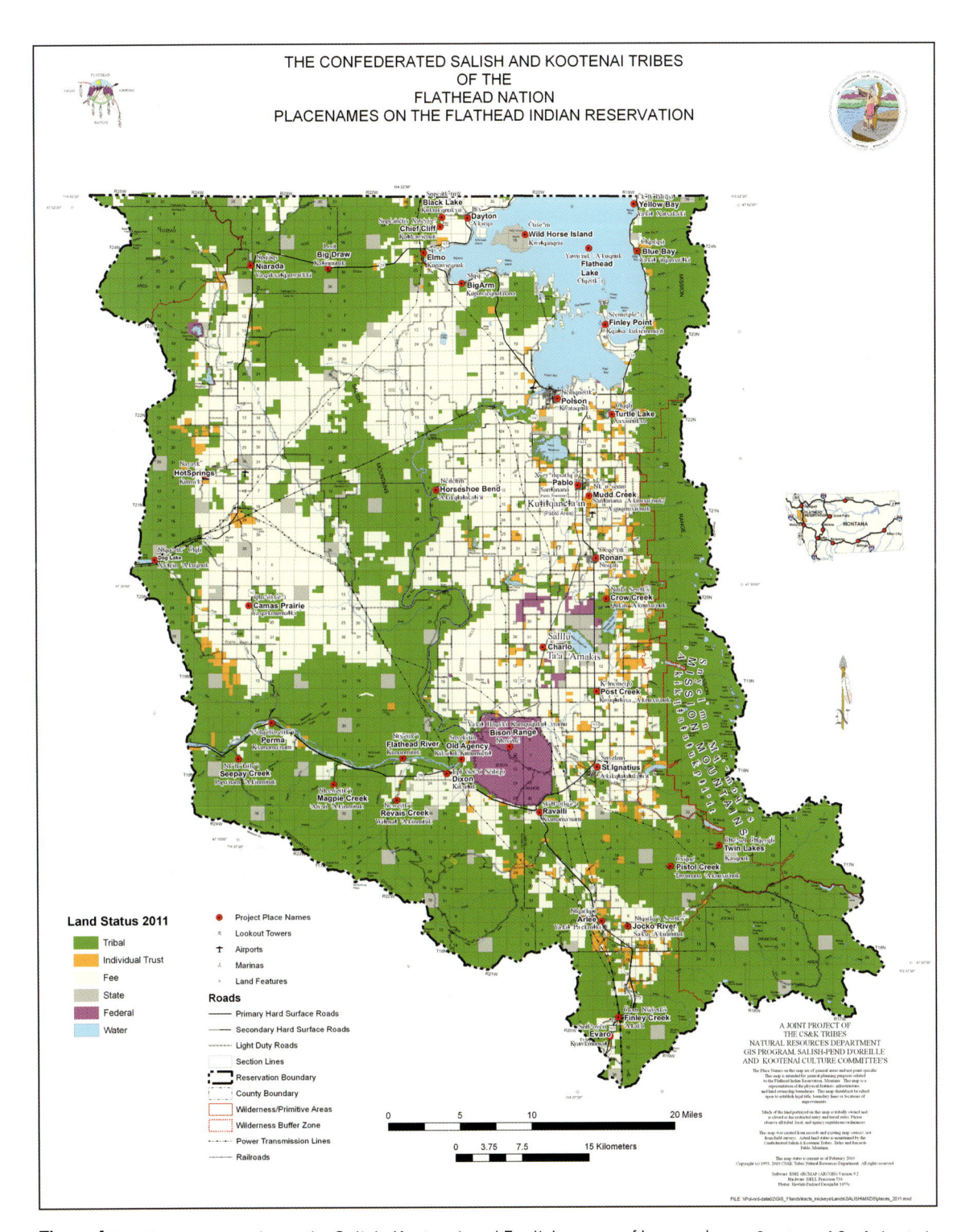

Chapter 3: Preserving tribal culture and history | 37

Figure 4. Land status map shows the Salish, Kootenai, and English names of known places. Courtesy of Confederated Salish & Kootenai Tribes

Reservation was sold as surplus land. Today through the efforts of the CSKT government, more than 60 percent of the total land is back in the possession of the Tribe and its citizens.

The CSKT began using GIS in 1988, and the CSKT GIS Program was established in 1990.

The Salish language is currently spoken by fewer than fifty people, most of whom are over seventy years old. There are no first language fluent Salish speakers under fifty years of age. Because of this, the Tribal Council of the Confederated Salish and Kootenai Tribes developed opportunities for imparting tribal traditions and values in each tribal department. The Natural Resources Department/GIS Department decided to make a land status map depicting known places with the traditional languages. With this particular project, the organizers wanted to create a map that would have the Salish, Kootenai, and English names of specific locations on the map. They also wanted to have the translations of what the Salish and Kootenai words mean if available. They were careful to make sure that there were no conflicts with the culture committees on the sensitivity of these names and locations.

These locations include areas that are commonly known to tribal members, as well as the towns on the reservation. The Elders' committee made the final decision on which names and locations the organizers could use for the final product that will be incorporated on the land status maps.

This project was developed to provide opportunities for both employee and public education on some of the traditional names used by tribal members to identify specific areas across the reservation in both Salish and Kootenai languages. Names and their translations were documented and placed into a database. By creating a georeferenced feature dataset containing the selected points and a geodatabase with the names and their translations, places could be mapped and displayed in the Salish and Kootenai languages.

The Tribal Council of the Confederated Salish and Kootenai Tribes has developed opportunities for imparting tribal traditions and values in each tribal department. This project is a direct result of creating an educational tool and a step forward in reaching that goal. Many departments within the tribal organization, as well as some of the education departments, are displaying these maps.

Winnemem Wintu—GIS helps resolve historical territory questions

Anne McTavish, MA, Geography, San Francisco State University; Mark Franco, Headman, Winnemem Wintu Tribe; Jerry Davis, PhD, Professor of Geography, San Francisco State University
San Francisco, California

When research requires using historical data, information specific to the issue at hand may need to be compiled, even inferred, from multiple sources. This is especially true regarding maps. Since major funding for collection and distribution of cartographic data and products was dominated by the state and major map publishers, maps were often produced to fulfill several needs at once: administrative, economic development, and general reference (Harley 2001). Accurate, detailed maps of historical Native territories are limited, and those that exist may be modified when reexamined for spatial accuracy. With computers, the Internet, and GIS software becoming more accessible, indigenous People and scholars around the world are reexamining existing maps or creating maps from historical data (Heasley 2003; Johnson, Louis, and Pramono 2005; Pearce 2004). Using contemporary GIS tools and techniques, a geodatabase may be developed to bring together information layers from historical paper maps, tabular descriptions, and contemporary GIS datasets, which may be compared once they are set to a common projection and used to compile additional data layers.

The Winnemem Wintu Tribe was historically located on the McCloud River and lived there until the 1940s, when they were forced to relocate because their villages were about to be inundated by Shasta Lake as Shasta Dam was completed in Northern California. Today their village is located in Redding. Because they are not federally recognized, they are incorporated as a nonprofit organization with 150 members. They remain active in their historical territory through spiritual ceremonies; participation in meetings that affect the McCloud River with Pacific Gas & Electric (PG&E), the Bureau of Reclamation, and CalTrans; and political actions to regain federal recognition.

The Tribe relies heavily on maps when discussing land-use changes with the various agencies. The Winnemem Wintu consider their historical territory to extend to Mount Shasta, but the north boundary has been considered controversial by researchers because Alfred Kroeber and C. Hart Merriam did not agree on where the line should be drawn (Theodoratus Cultural Research 1981). Because the Winnemem Wintu are not familiar with ArcGIS software, they rely on maps produced by agencies, researchers, and students. The Tribe maintains a collection of finished map files as paper documents, digital scans, or PDF files. The controversy between Kroeber and Merriam has been explored (McTavish 2010), and the techniques used to reexamine the Wintu boundaries are presented here as a case study.

European presence in California was primarily confined to the coast until the nineteenth century. There were no early systematic efforts to learn about the Indians in California and very few ethnographic records of any Tribe. Robert Heizer provided a summary of the limited statewide research and attempts to plot the distribution of the all the Tribes of California in *Languages, Territories, and Names of California Indian Tribes* (Heizer 1966). On a more local scale, Dorothea Theodoratus provided a summary of the early research done regarding the Wintu area, in *Native American Cultural Overview*, a document prepared for the Shasta-Trinity National Forest (Theodoratus Cultural Research 1981). Stephen Powers produced the first systematic map of the tribal areas in California in the 1870s, almost thirty years after most Tribes had been moved from their ancestral lands (Cook 1943).

In the early twentieth century, Alfred Kroeber and C. Hart Merriam differed on the location of the northern border of the Wintu because they disagreed about whether the Okwanuchu were part of the *Win-tūn* or the *Shastan* speaking group. Maps of the Wintu territory drawn by Alfred Kroeber (Kroeber 1925), Cora DuBois (DuBois 1935), and Frank LaPena (LaPena 1978, LaPena 2002) attributed the Okwanuchu area—just south of Mount Shasta—to the *Shastan* speaking group. However, the map drawn by Stephen Powers (Powers 1877), and the boundary descriptions given by Norel-Putis (Curtin 1889) and C. Hart Merriam (Merriam 1955), located the northern boundary of the *Win-tūn* speaking group at Mount Shasta. Examining the original writings of Kroeber, Merriam, and others sheds light on the way each inferred the boundaries they advocated.

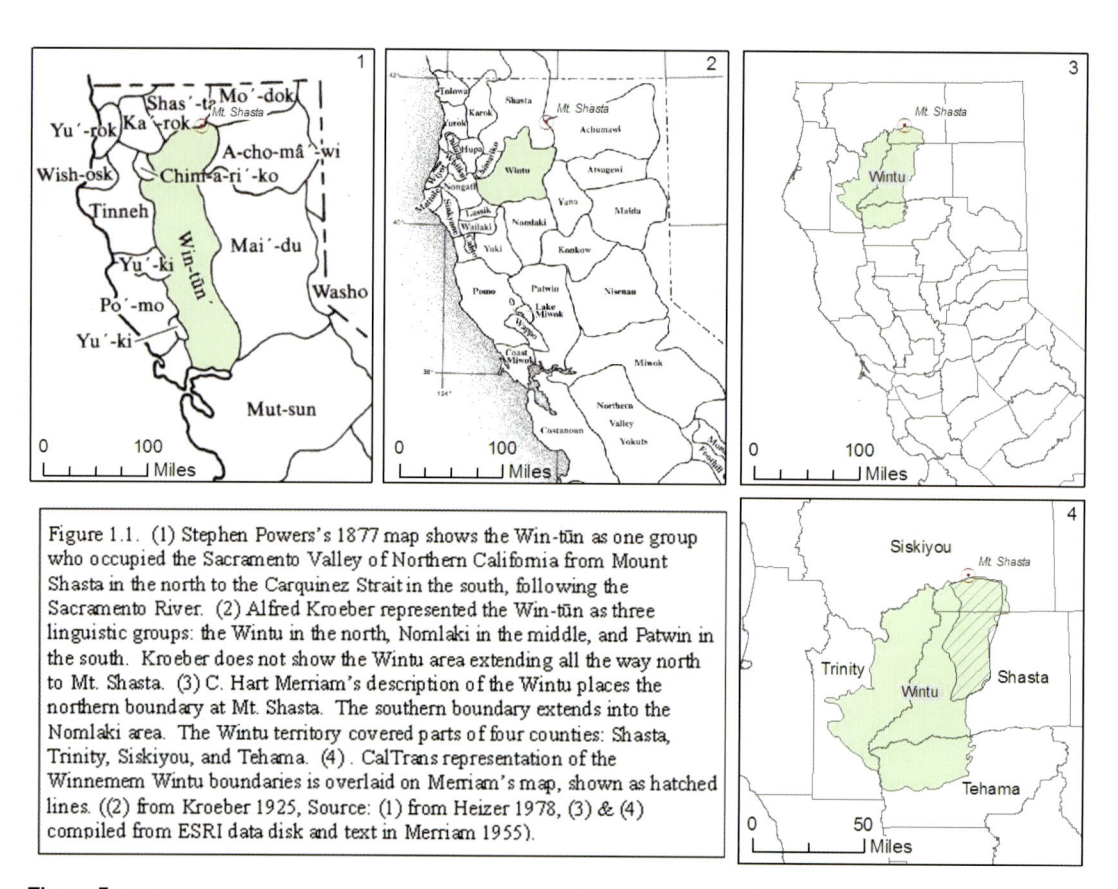

Figure 1.1. (1) Stephen Powers's 1877 map shows the Win-tūn as one group who occupied the Sacramento Valley of Northern California from Mount Shasta in the north to the Carquinez Strait in the south, following the Sacramento River. (2) Alfred Kroeber represented the Win-tūn as three linguistic groups: the Wintu in the north, Nomlaki in the middle, and Patwin in the south. Kroeber does not show the Wintu area extending all the way north to Mt. Shasta. (3) C. Hart Merriam's description of the Wintu places the northern boundary at Mt. Shasta. The southern boundary extends into the Nomlaki area. The Wintu territory covered parts of four counties: Shasta, Trinity, Siskiyou, and Tehama. (4). CalTrans representation of the Winnemem Wintu boundaries is overlaid on Merriam's map, shown as hatched lines. ((2) from Kroeber 1925, Source: (1) from Heizer 1978, (3) & (4) compiled from ESRI data disk and text in Merriam 1955).

Figure 5. Various map representations of the Winnemem Wintu territory in Northern California. Courtesy of Anne McTavish, San Francisco State University

Critical cartography provides a framework through which research-based analysis and interpretation of historical and spatial data may delve into the political aspects of maps, exploring how they exercise power, shape identity, construct knowledge, and promote social change (Crampton and Krygier 2005). Since the present circumstances of the Winnemem Wintu are derived from historical conditions, by critiquing the historical context actions that put the Winnemem Wintu on and off the map, it is possible to reveal and tease out attitudes that may have contributed to the boundaries advocated by various cartographers.

Alfred Kroeber was at the University of California in Berkeley for forty-five years, from 1901 until his retirement in 1946. In the early 1900s, Kroeber began to collect and edit the material for *Handbook of North American Indians*, which was finally published in 1925. As a linguist, Kroeber specialized in delineating the distribution of Indians in relation to language families. Neither Kroeber nor any of the contributors to the 1925 *Handbook* were very knowledgeable about the *Win-tūn* ethnography or languages. "Wintun [*Win-tūn*] speech is very imperfectly known, and its ramifications have been determined only in the rough" (Kroeber 1925). Kroeber and Roland B. Dixon, who was affiliated with the American Museum of Natural History, had researched the Shasta Indians and recognized six *Shastan* languages instead

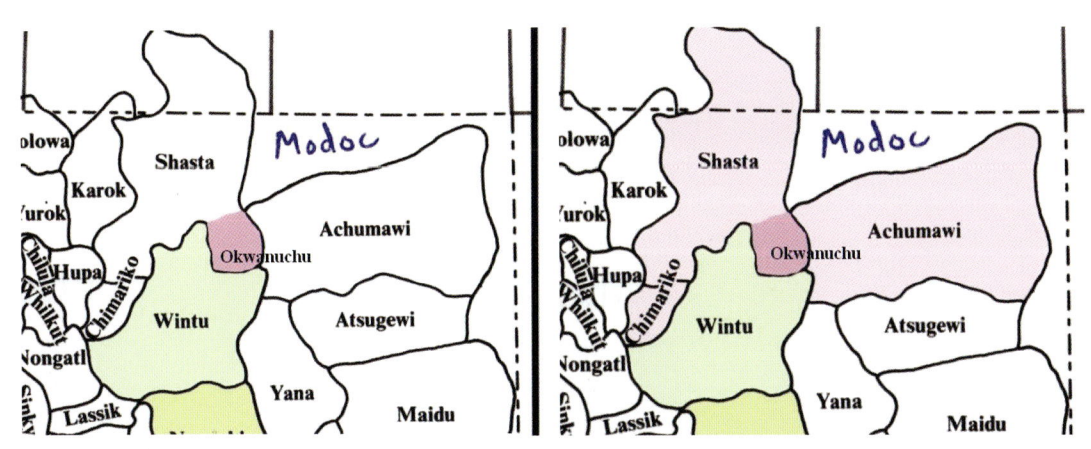

Figure 6. Okwanuchu. (1) Kroeber changed the northern border of the Wintu from Powers's map because he considered the extinct Okwanuchu to be part of the Shasta Tribe instead of Wintu. (2) The Okwanuchu provided a bridge on the map between the Shasta and Achumawi Tribes that supported Kroeber's and Dixon's recognition that the languages were related (Kroeber 1925). Courtesy of Anne McTavish, San Francisco State University (section of Kroeber's 1925 map; color and "Okwanuchu" added).

of the two recognized by Powers. Kroeber considered a group known as the Okwanuchu to be the missing link between the Shasta and Achomawi and represented them as part of the Shasta on his linguistic distribution map (figure 6). Explaining the changes from Powers's 1877 map, Kroeber wrote:

> *The reason for the long ignoring of the three languages adjacent to the Shasta is simple: no vocabularies were recorded, the tribes being numerically insignificant, and in one case on the verge of extinction when the white man came to northern California. Now they have dwindled so far—in fact to all practical purposes perished—that when we are hungry for any bits of information that would help to untangle the obscure history of these remnants of what may once have been greater peoples, we must content ourselves with brief, broken vocabularies and some general statements about their speakers obtained from the neighboring nations.* (Kroeber 1925)

Another scholar cited as an expert on California Indians is Merriam. His work was largely unpublished at the time of his death in 1942. In 1955, Merriam's description of the Wintu boundaries was published, unaltered, in *Studies of California Indians*. Merriam's description was used to locate the boundaries on the map using GIS data for watershed boundaries or rivers, as shown in figure 7. Once mapped, Merriam's boundaries could be compared to other georeferenced maps prepared by Powers, Kroeber, DuBois, LaPena in 1978 and LaPena in 2002. Comparing the maps created by Powers and described by Merriam shows similar northern boundaries for the Wintu. Knowing that his view differed from Kroeber's and others who attributed the Okwanuchu area to the Shasta Indians instead of the Wintu, Merriam explained that *Okwanuchu* was a directional word used by the Shasta Indians that meant "south of here."

At various times from 1884 through 1889, Jeremiah Curtin, a linguist working for the Smithsonian, interviewed a number of Wintu, including Norel-Putis. Norel-Putis had broader experiences than many Wintu. He was an honored person, probably a chief, and well-traveled in the Wintu area. He was reported

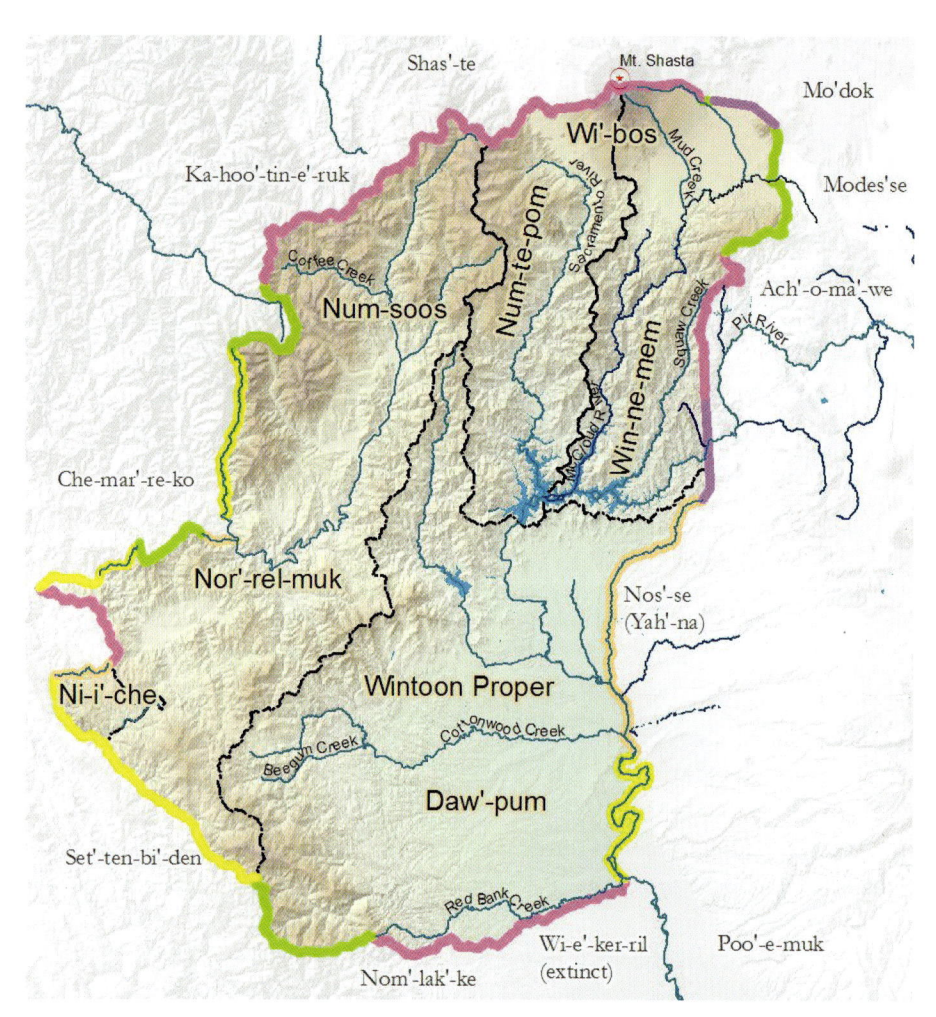

Figure 7. C. Hart Merriam's Wintu boundaries. These were described in *Studies of California Indians*, published by the University of California Press through the efforts of Robert Heizer. Since Merriam had not created a map from this description, the boundaries for this study were located using rivers and watershed boundaries (Merriam 1955). Courtesy of Anne McTavish, San Francisco State University. Compiled from Merriam's text description, Cal-Atlas data, and Esri Data & Maps

to have been 100 when he died on March 12, 1894, so he would have already been an adult before the Wintu experienced their first contact with whites (Dotta 1980). As an adult, his reports are considered "contemporaneous," meaning the villages he described were inhabited all at the same time and in his lifetime (Dotta 1980).

In 1888, Norel-Putis and others of the Wintu and Yana Tribes asked Curtin to write a letter for them to the president of the United States, Benjamin Harrison, which Curtin agreed to do in order to:

> ... tell the President what a homeless condition they were in, how the white men drive them from place to place. I told them to find out how many of each tribe were living and draw up a paper

stating their condition, and I would try and do something for them in Washington. (Curtin 1940 in Guilford-Kardell 1980)

There isn't a map associated with the Wintu-Yana petition, but Norel-Putis described the *Win-tūn* territory:

The Wintu people before the Whites came into the land of our fathers owned and inhabited the country extending from Mount Shasta on the north to Carquines [sic] Straits on the south. The western boundary of this country was the mountain range west of the Sacramento Valley except in the region between north Yallo Valley and Edgewood where the line went west of the range and the Wintu occupied one-half of the Trinity. (Curtin 1889)

GIS methods

Historical descriptions often referred to places, but the research required making maps that had to be compiled from many sources. Using ArcGIS from Esri to employ a geodatabase to bring together information layers from existing paper maps, georeferenced historical descriptions, and GIS datasets (all set to the same projection), the various inputs could be examined together, making it easy to see patterns and spatial relationships. These patterns made it possible to draw new insights from past events that were not evident when examining the data sources separately. Digital data sources available from the Internet and Esri provided spatially consistent base layers that were built upon georeferenced maps and newly compiled data layers. Paper maps were used to the extent they could be found. Text descriptions were the only source of spatial information for two references. The data collection and processing necessary to complete this study and the resulting maps were conducted over a period of three years, with research being conducted at various libraries, including the Shasta Historical Society, College of the Siskiyous, Redding Public Library, and various libraries at UC Berkeley, including the Bancroft. Research was also conducted at the National Archive Record Administration (NARA) in San Bruno, with records from the USDA Forest Service in Mount Shasta and US Bureau of Reclamation, and with records collected by the Winnemem Wintu. ArcGIS was used to store and manage the geospatial data, including publicly available basemaps, and to create maps showing the disputed territory and historical boundaries. With the analysis detailed on a poster, the Winnemem Wintu may now clearly explain to others the source of the controversy and the validity of their historical claim that they are the ones spiritually responsible for the contested northern area.

References

Cal-Atlas Geospatial Clearinghouse. 2007. http://www.atlas.ca.gov/. ArcGIS 9.3.1. 2009. Esri, Redlands, California, Esri Data & Maps 2009. Esri, Redlands, California.

Land Survey Information System. 2009. Bureau of Land Management in partnership with the US Forest Service. http://www.geocommunicator.gov/LSIS6/map.jsp.

Cook, S. F. 1943. *The Conflict Between the California Indian and White Civilization*. Berkeley and Los Angeles: University of California Press.

Crampton, J. W., and J. Krygier. 2005. "An Introduction to Critical Cartography." *ACME: An International E-Journal for Critical Geographies*, 4, 11-13.

Curtin, J. 1889. "Wintu and Yana Petition to Great Chief Benjamin Harrison, President of the United States." Winnemem Wintu Library, Redding: Office of Indian Affairs, Washington, DC.

Dotta, J. 1980. "Some Elements of Wintu Social Organization as Suggested by Curtin's 1884-1889 Notes." In *Papers on Wintu Ethnography: 239 Wintu Villages in Shasta County Circa 1850*, edited by M. Guilford-Kardell, 118-29. Redding, California: Occasional Papers of the Redding Museum and Art Center, No. 1.

DuBois, C. 1935. "Wintu Ethnography." In *American Archaeology and Ethnology*, edited by A. L. Kroeber, R. H. Lowie, and R. L. Olson, 1-148. Berkeley: University of California Press.

Guilford-Kardell, M. 1980. "Papers on Wintu Ethnography: 239 Wintu Villages in Shasta County Circa 1850." 131. Redding Museum and Art Center, Redding, California. Occasional Papers of the Redding Musuem No. 1.

Harley, J. B. 2001. *The New Nature of Maps: Essays in the History of Cartography*. Baltimore, Maryland: Johns Hopkins University Press; published in cooperation with the Center for American Places.

Heasley, L. 2003. "Shifting Boundaries on a Wisconsin Landscape: Can GIS Help Historians Tell a Complicated Story?" *Human Ecology*, 31, 183-213.

Heizer, R. F. 1966. *Languages, Territories, and Names of California Indian Tribes*. Berkeley: University of California Press.

———. 1978. "Introduction." In *Handbook of North American Indians, Volume 8, California*, Sturtevant, William C., general editor, and Robert F. Heizer, volume editor, 1-5. Washington, D.C.: Smithsonian Institution.

Johnson, J. T., R. P. Louis, and A. H. Pramono. 2005. Facing the Future: Encouraging Critical Cartographic Literacies in Indigenous Communities. *ACME: An International E-Journal of Critical Geography*, 4.

Kroeber, A. L. 1925. *Handbook of the Indians of California*. Berkeley: California Book Company.

LaPena, F. 1978. "Wintu." In *Handbook of North American Indians, Volume 8, California*, Sturtevant, William C., general editor, and Robert F. Heizer, volume editor, 324-40. Washington, D.C.: Smithsonian Institution.

———. 2002. "Wintu Tribal Territory." In *Journey to Justice: The Wintu People and the Salmon*, edited by A. R. Hoveman. Redding, California: Turtle Bay Exploration Park.

McTavish, A. 2010. "The Role of Critical Cartography in Environmental Justice: Land-Use Conflict at Shasta Dam, California." Master's thesis, San Francisco State University.

Merriam, C. H. 1955. "Tribes of Wintoon Stock." In *Studies of California Indians*, edited by the Staff of the Department of Anthropology of the University of California, 3-25. Berkeley and Los Angeles: University of California Press.

Pearce, M. W. 2004. "Encroachment by Word, Axis, and Tree: Mapping Techniques from the Colonization of New England." *Cartographic Perspectives*, 48, 24-38.

Powers, S. 1877. "Tribes of California." In *Contributions to North American Ethnology*, vol. 3, edited by J. W. Powell, 229-42, 518-33. Washington, D.C.: Department of the Interior. US Geographical and Geological Survey of the Rocky Mountain Region. J. W. Powell, director.

Theodoratus Cultural Research. 1981. In *Native American Cultural Overview*, edited by P. I. Dorothea Theodoratus. Redding, California: USDA Forest Service, Shasta-Trinity National Forest.

Chickasaw Nation: Using GIS for historical and cultural preservation and awareness

John Ellis, Director, Department of Geospatial Information, the Chickasaw Nation
Ada, Oklahoma

The Chickasaw Nation's rich history involves two regional homelands. The first of these homelands was centered near Tupelo, Mississippi, while the second is centered in Tishomingo, Oklahoma. Prior to Indian removal in the early 1800s, the Chickasaw People enjoyed great prosperity in the southeastern United States. The Chickasaws were a highly developed society with a complex system of government, a mostly agrarian lifestyle, and strong cultural identity.

After the Chickasaws reluctantly ceded their lands east of the Mississippi to the US government over a period of forty-six years, the federal government achieved its goal of removing the Chickasaws to Oklahoma, then known as "Indian Territory." Removal from the original homelands, which began in 1837, resulted in the Chickasaw People establishing their new homelands in south-central Oklahoma.

Upon arrival in Oklahoma, the Chickasaw People were recognized as part of the Choctaw Nation, but in 1856, the Chickasaw People reestablished the Chickasaw Nation as a sovereign government. The government was organized, a constitution was adopted, and the capital was established in Tishomingo, Oklahoma.

Statehood for Oklahoma came in 1907 and brought federal control over the operation of the Chickasaw Nation Government. In 1970, Congress passed legislation that allowed for the election of tribal leaders once again. In 1983, the Chickasaw Nation adopted its current constitution. The Chickasaw Nation's boundaries encompass 7,443 square miles in south central Oklahoma, covering all or parts of thirteen counties.

Because preserving cultural identity is a primary concern, the Chickasaw Nation is using GIS as part of an ongoing project to ensure cultural preservation of artifacts that might otherwise be lost due to urbanization. By locating sites of historic importance among modern day surroundings in Oklahoma and in the original homelands area of northern Mississippi, artifacts can be identified and repatriated through NAGPRA.

This project is intended to locate these sites of historic importance through the overlaying of GIS layers over topographic and orthographic images. One of the challenges of this project is the dissemination of accurate historical data to the users in a timely fashion. Another challenge is providing the user with all the resources needed to make informed decisions in their preservation endeavors.

Historical information is gathered from various departments and used by the Department of GeoSpatial Information (GSI) to delineate feature classes and create raster datasets. Because of the vast amount of historical and archeological data that needs to be created and edited, a GIS technician was assigned to work with various historical and cultural departments to set up a GIS framework for preservation. This framework consists of a database that stores feature classes and raster datasets that have been acquired or created. Information is then distributed through maps and presentations to the requesting parties. A website was created to handle the influx of data requests. The site contains historical cession areas by year, 1830 Cession Surveys, Public Lands Survey System (PLSS) data, parcel data, street maps, counties, elevation (hillshade) data, topographic maps, aerial photography (from the North American Imagery Program and Pictometry), and satellite imagery. The user can search for data and create simple maps, allowing a GIS technician to work on more complex projects. This process allows the users more freedom to conduct their own research, while having the guidance of the GeoSpatial Information Department when needed. Website and database development are ongoing projects with data being added when information is provided to the department.

The process began with creating a Mississippi database to store all original homelands data. The Tribe's Division of History and Culture acquired copies of the 1830 US government homeland surveys that displayed important roadways, footpaths, and land holdings. These 1830 Cession Surveys form the basis from which all historical Mississippi information is delineated. The surveys were loaned to the GSI department to be converted into a digital format. The surveys were scanned using an HP Design Jet 4500 and stored in a TIFF format. Then using Adobe Photoshop CS4, the TIFF files were cropped to remove excess text and titles, leaving only the grid and survey details. The next step was to georeference the cropped surveys to a PLSS grid that was downloaded from the Mississippi Automated Resource Information System (MARIS). Upon completion of the georeferencing process, the roadways, footpaths, and land holdings displayed in the surveys were digitized into the appropriate point, line, and polygon shapefiles.

Another Mississippi project currently underway involves digitizing archeological sites. The Chickasaw Nation received a collection of twenty, seven-and-a-half minute topographic quad maps of the original homelands area from the Mississippi Department of Archives and History. These maps have archeological sites of interest highlighted. The topographic quad maps were scanned using an HP Design Jet 4500 and stored in a TIFF format. They were then georeferenced using the federal PLSS downloaded from MARIS. A polygon shapefile was created and the sites highlighted on the maps were then digitized.

While the primary focus has been the original homelands of Mississippi, a need to provide information for historic Chickasaw sites of importance in Oklahoma has become apparent. The Division of History and Culture has created a list of sites that were deemed significant, and a GPS unit has been used to record the areas. This information was then exported into a polygon shapefile. The locations are categorized as historic churches, schools, and sites. Also available are historical routes and railroads from 1889, which were digitized from the *Historical Atlas of Oklahoma* and then georeferenced to Esri state data.

The 1830 Mississippi Cession Surveys were given to the GSI department in early 2006. With a staff of just three employees, each person was in charge of multiple projects. Projects were completed in order of importance and need, causing the more laborious, time-consuming projects to be put on hold. Over the last four years, the staff has experienced major growth, more than tripling in size. With the staff now at thirteen, projects once put on hold are dispersed by subject matter to individuals with expertise in that topic. In 2008, the cultural and historical preservation projects were assigned to a GIS technician. The 1830 Mississippi Cession Surveys were finished in October 2009. Work is currently being done on the topographic quad maps with highlighted archeological sites in Mississippi, which were received in the spring of 2008. All of the Oklahoma data that is available is fully completed, but some of the historical site GPS points have been found to be incorrect and will be re-shot in the future by the Division of History and Culture. These projects have been implemented in a time frame of about four years, with a large majority of the project taking place during the last two years. The key has been to finally have a technician that can spend a dedicated portion of his time working solely on history and culture projects.

All of these projects had challenges, but three projects became particularly problematic. The 1830 Cession Surveys that were received were copies of the originals, and as such, many of the copies were of poor quality with dirty or large black areas where information could not be seen. When the surveys were cropped in Adobe Photoshop, they had to be cleaned up using the filters provided in that program as best as possible. Most of the surveys that did have problems were made legible, but some of them were in such poor condition they could not be cleaned. Yet even using the imperfect surveys was of great use because vast amounts of information were provided. Another challenge dealt not with the surveys themselves, but with the notes that the surveyors had written about them in the 1830s. The notes provide a large amount of information and actually tell about the location of sites that are not found on the surveys themselves, but the writing style is unlike that in use today. The writing is very beautiful to look at but almost impossible to read. The third challenge was the aforementioned lack of dedicated time to the survey project.

The Chickasaw Nation historical website has been of great use. Maps can now be made to display all of the historical information that has been georeferenced and digitized. The maps may include topography or hillshades, or they may show historic trails or removal period ownership. A slide presentation documenting the georeferencing process and possible uses of the 1830 surveys has also been created. Additionally, there are some photos of GPS training and collection from Chickasaw cemeteries. All the resulting information (data, maps, presentations, and so on) helps preserve the past of the Chickasaw People by allowing these historic sites to overlay modern-day maps of Mississippi. Preservation is a primary focus for the Chickasaw People.

The historical and cultural preservation project is ongoing. As archeological sites and historical documents reveal new findings, geospatial data creation, editing, and processing continues. The topographic maps, which contain highlighted sites of interest, show more than 800 sites and contain many more yet to be digitized. The 1830 Cession Surveys contain 320 individual surveys and will contain many more

Figure 8. The Chickasaw Nation's new Cultural Center. Courtesy of Chickasaw Nation

as information becomes available. One hundred and seventy-five roads and pathways and 145 sites of interest have been digitized from the 1830 surveys. The outcome of these projects has been tremendous. Historical site and road information can be instantly overlaid onto modern day topographic maps and aerial imagery. When new highways or urban development projects are proposed, the historical information can be combined with modern GPS information to see if an archeological site will be disturbed or destroyed.

The Chickasaw Cultural Center, located on 109 acres in the northeast corner of the Chickasaw National Recreation Area in Sulphur, Oklahoma, opened to the public in the summer of 2010. Since that time, it has averaged more than 1,000 visitors per week. The Cultural Center features an exhibit center, art gallery, the Chickasaw Hall of Fame Honor Garden, large-format theatre, amphitheatre, sky terrace, and a traditional Chickasaw Village. The center is a welcoming home for all to come and learn more about the heritage and traditions of the Chickasaw People.

Economic development on tribal lands

In the United States, Tribes have unique sovereign rights over their lands. With these rights come responsibilities to manage tribal lands in an appropriate and sustainable manner for the benefit of tribal members. Without clear and precise definitions of tribal land boundaries, managing and protecting these lands is challenging if not impossible. Few Tribes and indigenous communities have jurisdiction over all of their usual and accustomed lands. In many cases, original treaties formed with the federal government were revised through subsequent treaties or legislation, which diminished or facilitated the sale of tribal lands. This has resulted in complex tapestries of land ownership within many reservation boundaries. GIS provides a framework for not only managing the boundaries of tribal lands but the tribal ownership rights within those lands, which are often complex. A single tribal parcel or property may have large numbers of landowners. Tribes also have the ability to move lands into a unique federal trust status, which offers specific privileges and exemptions from taxation. Transactions that move tribal lands into trust, and involve the acquisition or sale of tribal lands, all require extensive spatial review throughout an often complex process involving tribal members, the tribal government, and the federal government. Spatial data representing the parcels and boundaries involved is at the heart of these transactions. GIS naturally lends itself to realty transactions and is used across governments of many types for this purpose around the world.

Managing realty within a tribal government, however, involves more than knowing where tribal lands are and understanding their ownership. Increasingly Tribes are actively purchasing new lands to restore access to lands and resources diminished over the years. The health of tribal communities often depends on the economic opportunities available to tribal members. Making sound economic decisions relies on an acute understanding of what lands the Tribe does not wish to disturb in order to protect and sustain its unique cultural identity. For these reasons, real estate and economic development are closely related. GIS is an important tool for both because it allows stakeholders to greater understand current tribal land use and make better decisions about future land use.

The stories in this chapter are about five tribal governments that applied GIS to their real estate programs and have realized a variety of benefits by doing so. They are the Fond du Lac Band of the Lake Superior Chippewa, the Agua Caliente Band of Cahuilla Indians, the Confederated Tribes of Siletz Indians, the Lummi Nation, and the Navajo Nation.

Strategic land-use planning with GIS

Tim Krohn, GIS Specialist, Fond du Lac Band of the Lake Superior Chippewa
Cloquet, Minnesota

Land has a deep connection to one's heritage. A comprehensive use of GIS is extremely helpful in maintaining ownership of this heritage. Over 600 years ago, the Anishinabe (Ojibwe/Chippewa) lived on the shores of the Atlantic Ocean near the mouth of the Saint Lawrence River. Over time the Tribe migrated to Lake Superior and split into two communities—one north and one south of the lake. In the 1680s, the ancestors of the Fond du Lac People arrived at Madeline Island north of today's Ashland, Wisconsin. The Anishinabe traded with and on occasion fought with and against the Dakota Tribes, who inhabited the western shores of Lake Superior and lands farther inland. In the early to mid-1700s, several encampments were established by the Anishinabe and French in the area of the mouth of the Saint Louis River at the west end of Lake Superior. It was here that a fur trading camp was established and called Fond du Lac.

At the Treaty of September 1854, a large area of nearly 6.3 million acres was ceded to the United States with a 125,000-acre reservation set aside for the Fond du Lac People. Due to misconceptions of geography during the treaty negotiations, the reservation was surveyed twice, first in 1858 and then in 1860. The very important rice lakes for the Fond du Lac People were outside the reservation as surveyed based on the initial treaty description of the boundary. Two months after the survey was completed, letters were sent to Washington complaining about the absence of the rice lakes inside the reservation boundary. The second survey added in the rice lakes but reduced the reservation size to just over 100,000 acres. The American government justified this reduction by stating that the Fond du Lacers were getting better land, a swap of value for value, not acre for acre, and that 100,000 acres was closer to the size contemplated by the stipulations of the treaty.

The Nelson Act of 1889 changed the land ownership of the reservation from wholly Band-owned to individual ownership under trust, known as allotments, and reduced Fond du Lac land ownership

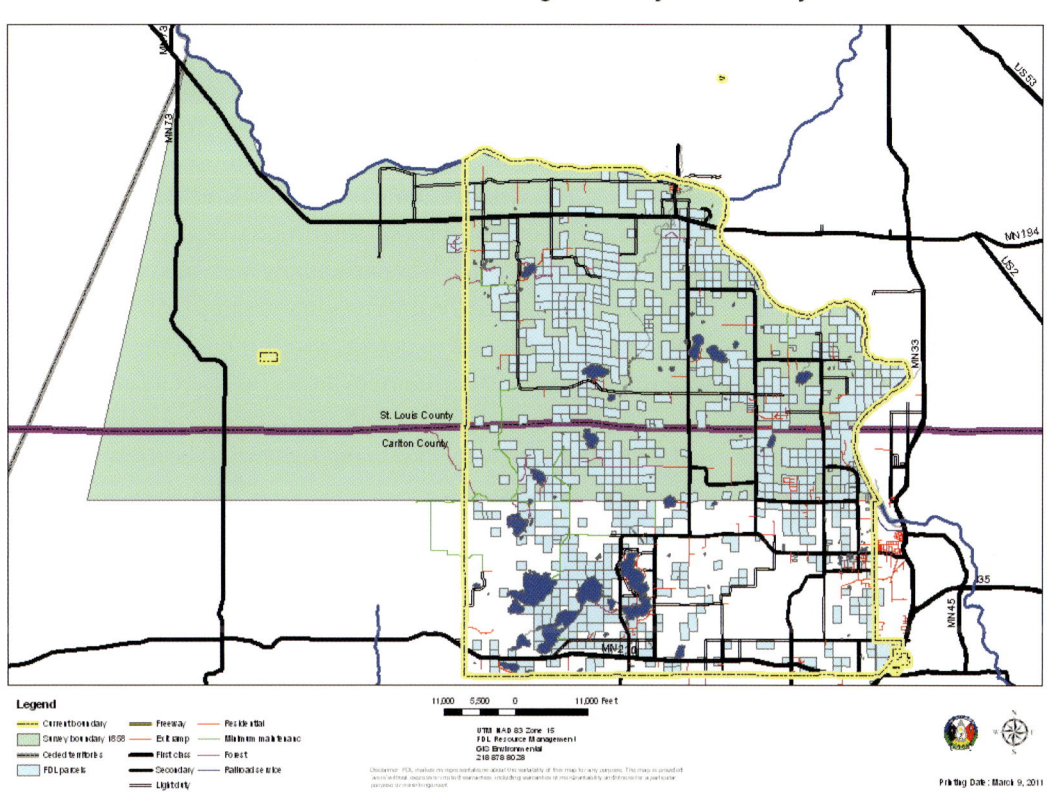

Chapter 4: Economic development on tribal lands | 51

Fond du Lac Boundaries - Original Survey and Resurvey

Figure 1. Surveys of treaty lands of the Fond du Lac. Courtesy of Fond du Lac Band of the Lake Superior Chippewa

even further to about 30,000 acres. The Nelson Act also opened up the land to settlement by nontribal interests. In the late 1910s, there was a push to convert individual allotments to fee land subject to property taxes. Eventually this reduced Fond du Lac ownership to only 9,000 acres. Land ownership under trust became highly fractionated with most parcels having 50 to 200 owners and one parcel having over 1,800 owners. This fractionation has become a very large problem for the Tribe in managing these lands. Through a concerted effort to recover their lands, today Fond du Lac has over 30,000 acres under trust owned by the Fond du Lac Band, the Minnesota Chippewa Tribe, or individual tribal members.

In the 1854 treaty, which established the Fond du Lac Reservation, the description of the reservation begins at an island in the Saint Louis River called *Paw-paw-sco-me-me-tig*. For many years the location of the island mentioned in the treaty was unclear. However, recently by incorporating a variety of historical records and documents into a GIS, the approximate location of the island has been discovered. This was accomplished by incorporating spatial data from historical documents into ArcGIS, including the metes and bounds surveys of the reservations done in 1858 and 1860, as well as letters and hand drawings from the mid-to-late 1800s. The island is also mentioned in Canadian surveyor David Thompson's survey notes from the 1820s. It appears that on July 26, 1825, Thompson and crew camped overnight on this island, which is the only reference to camping in his survey notes along the Saint Louis River. This historical reference was incorporated into a GIS project using ArcGIS Desktop. The approximate location of the island was then verified in the field using a canoe and a handheld GPS.

In 1997, Fond du Lac began applying GIS in support of a strategic land-use plan for the Band that involved mapping roads, water, and ownership within the reservation boundary. Prior to deploying GIS, Fond du Lac did not have a comprehensive catalog of its land, resources, and homes. Due to the mosaic nature of land ownership within the Fond du Lac Reservation—private fee land, public lands, tribal trust, Band trust, and trust allotment—it was sometimes hard to determine who owned what. Due to the fractionated nature of ownership of lands under trust, it was particularly hard to identify ownership and percent ownership for a particular parcel. Also, since the data was maintained off the reservation at the Bureau of Indian Affairs (BIA), it was additionally hard to get and use this data. Through the use of GIS, these issues have been resolved, and the Band now has immediate access to realty and supporting information. Additionally, through applying GIS as a common unified framework for the tribal lands, historical mistakes were identified in the title of land ownership in the 1920s that the Tribe is now working to correct.

The use of GIS by the Band has evolved since the 1990s, with the GIS now supporting a wide variety of topics, including parcel ownership mapping, home-lease planning, support for settling property boundary disputes, forest inventory, wildlife hunting zones, fisheries management, bathymetric and watershed management, wetland management and protection, and air emissions monitoring. While this work is coordinated by the GIS specialist, a number of Band Government staff apply GIS to their specialties, including wetlands, land, air quality, forestry, and water.

In order to have an accurate GIS system with adequate precision for the many uses described above, Fond du Lac has invested in extensive surveying and field data collection to achieve a near survey-grade accuracy for realty data. The initial GIS parcel layer, which was developed by a contractor, was incrementally improved by collecting coordinates for all of the government corners using a map-grade GPS, a survey-grade GPS, or prior surveys. Collecting 850 corner coordinates was a multi-year task. Many corner coordinates were collected in winter when snow and cold made travel over wetlands much easier. Once the corner coordinates were collected, sections were broken down into their proper forties or lots. The county auditor's data was downloaded into a database. Each parcel's legal description was compared to the recorded deed, and then each parcel was drawn as per the legal description on the recorded deed with any differences noted in the final file. Finally, parcel maps and data files were uploaded to Fond du Lac's website for all to see and use.

An accurate and up-to-date GIS system for managing realty at Fond du Lac has led to a variety of benefits. Accurate maps showing wetlands and property lines now help the Band leadership approve or disapprove land purchases and land leases. Within the Fond du Lac Reservation, and especially on trust land, there is a scarcity of buildable home sites. By using GIS to identify high ground suitable for septic systems, Fond du Lac can avoid purchasing or placing homes on unsuitable locations.

Sustaining economic development on sacred lands

Beckie Howell, GIS Manager, Agua Caliente Band of Cahuilla Indians
Palm Springs, California

Since time immemorial, the Agua Caliente Band of Cahuilla Indians inhabited and governed some 2,000 square miles of ancestral land in the Palm Springs area of California. The members of the Tribe lived in several well-established communities in the canyons and surrounding mountains and desert floor. The Agua Caliente Indian Reservation is located in Riverside County, California. It comprises approximately 31,500 acres, not including nearly 4,000 acres of off-reservation, tribal trust land. The reservation shares jurisdiction with the cities of Palm Springs, Cathedral City, Rancho Mirage, and unincorporated Riverside County. The Agua Caliente Tribal Administration Plaza is located in the city of Palm Springs.

The arrival of the railroad in the 1860s had a dramatic impact on the lives of tribal members, especially when the federal government gave all the odd-numbered sections of land in the Coachella Valley to the Southern Pacific Railroad. When President Ulysses S. Grant established the present Agua Caliente Indian Reservation by Executive Order in 1876, only the even-numbered sections were still available, thus creating the present reservation in a checkerboard pattern.

To encourage economic development, the federal government allocated the bulk of the reservation land to the individual members of the Tribe in a process called "allotment" that lasted until 1959. In the same year, and for the same purpose, Congress authorized only this Tribe and its members to lease their land for up to 99 years. Much of Palm Springs and adjacent Cathedral City, Rancho Mirage, and unincorporated Riverside County is built on such land allotted to individual tribal members and leased to developers in this way. However, under the Tribe's constitution, adopted in 1955, and federal law, the Tribe maintains primary control over the use and development of all land on its reservation, including those parcels included in cities located on the reservation.

The activities of the Agua Caliente Band of Cahuilla Indians have had a variety of economic impacts on the Coachella Valley, stemming not only from the last ten years of its casino operations, but also the extensive holdings remaining in the Tribe's historic land base, underlying much of what is now the cities of Palm Springs and Cathedral City, and portions of the city of Rancho Mirage and Riverside County. Today, much of the Tribe's original reservation of 31,500 acres has been highly developed into a modern desert tourism center, as well as extensive upscale retail and residential uses. The reservation includes billions of dollars worth of valuable real estate as well as invaluable desert and mountain habitats.

The Tribe began benefiting from GIS services in 1998 with support from the BIA state office located in Sacramento, California. However, these services became problematic due to the long turnaround time for maps. The Tribe needed access to timely and accurate geographic information to support important decision making, so they established their own in-house GIS capacity in 2002. The Tribal GIS Group is currently located at the tribal government offices and is housed within the Planning and Development Department. The Tribal GIS Group is funded in part by the tribal government, but also through various grants. Throughout the years, the Tribe has received grant funding through the BIA Aid to Tribal Government (ATTG) Grant and the Environmental Protection Agency (EPA) General Assistance Program (GAP) Grant. Having an onsite GIS staff allowed the tribal government to more efficiently manage

Figure 2. Economic development on the Agua Caliente tribal lands. Information and graphics provided by the Agua Caliente Band of Cahuilla Indians

the reservation land holdings and initiate a large number of time-sensitive projects and analysis that otherwise would not have been possible. This transformation has made GIS an invaluable resource to the tribal government.

Currently, the Tribe uses GIS for numerous projects, including environmental protection, real estate management, land use and planning, urban and economic development, cultural resource preservation, and construction. The Tribal GIS Group has aided in the completion of numerous strategic plans, which have been approved by the Tribe, including the Tribal Habitat Conservation Plan, Pre-Disaster Mitigation Plan, Fire and Forest Management Plans, and the Section 14 Master Plan.

Perhaps the most important way the Tribe uses GIS is to manage the land holdings of the Tribe and individual tribal members. The land ownership records, also known as the land status, are important tools used by the Tribe for decision making. It is critical that these records are accurate, complete, and current. Keeping the land ownership status on the reservation up-to-date is a great challenge for the Tribal GIS Group as well as finding ways to make the land status records readily available to members of the tribal government. Reservation land status is continuously changing as properties are bought, sold, leased, and inherited. All land owned by individual tribal members that is held in trust is managed by the BIA. This becomes problematic because the Tribe is often not made aware when land transactions among tribal members occur. The Tribal GIS Group often needs to research land transactions to identify changes in status. Another problem the Tribal GIS Group is attempting to overcome is the management of all pertinent documents related to each property. In many cases, it is necessary for tribal government team members to gain access to records of ownership, easements, and maps. These records are managed by the Tribal GIS Group and need to be made available for quick viewing, through the network, while maintaining appropriate security restrictions. The Tribal GIS Group is continually looking for new and more efficient ways to make this critical information available to the tribal government staff.

The evolution of the solution to keep tribal land status records up-to-date is an ongoing process. Currently the GIS manager has established standard operating procedures as to the methodology of updating the land status when changes have occurred, as well as ways to identify and verify these changes. The land status data layer was originally created using property descriptions from the BIA Land Title and Records Office and has been further refined using legal descriptions from grant deeds provided by Riverside County. The Tribe has also been awarded Bureau of Land Management (BLM) grants for cadastral resurveys of the reservation townships to better represent the data geographically. The research involved in updating the records includes close communication with the BIA, annual review of updated title status reports, examination of the Riverside County assessor's parcel data for indications of development and potential fee transfers, as well as visual inspection of aerial imagery for changes in land use. This is an ongoing process as land transactions occur on a continual basis. When land status changes have been identified and verified, the production database is updated, as are a series of critical maps and exhibits. Once the production database has been updated, the updates are pushed to the consumption database where the data is made available to users through the various end-user applications that serve the data. The GIS Group makes this information available to decision makers through a variety of resources including, maps, web applications, published ArcReader files, ArcMap files, and ArcGIS Explorer.

The GIS manager also acts as the database administrator and takes the lead on major mapping projects. She is responsible for managing the department budget, designing and overseeing the databases, and assigning tasks to the GIS staff. The GIS analyst handles all day-to-day mapping requests of the tribal organization, and also takes on long-term mapping and analytical projects. Tribal GIS data is stored in an enterprise geodatabase using SQL Server and ArcSDE. The data is stored in two databases, one for production, used by the GIS staff, and one for consumption, used by the GIS users. As data is updated, it is pushed from the production database to the consumption database, so the GIS users always have the most current data. Tribal GIS collected and cataloged documents for each property and stored them in digital format. A hyperlink field was created in the database table that provides the user with one-click access to a web page that contains links to all relevant documents in PDF format. Now users can view and print maps, as well as access property documentation all within the same GIS application.

In addition to a number of ArcGIS Desktop licenses, ArcGIS Server is used to create, manage, and serve the tribal web applications. The staff also uses the ArcGIS Server Image extension to manage and serve nearly a terabyte of raster data. By using ArcGIS Server to develop a land status web map project, reference and research was made more efficient. Previously, when projects were being researched or worked on, team members would have to collect the hard-copy files from a central filing location outside their office. Now, with online database access, team members can bring up all property documents available, with one click, right from their desks. This method is slowly phasing out the need and use for storing hard-copy documents, which saves resources and space. When the web mapping project was first initiated, the intent was to provide a service only to the planning staff, but as the project progressed, it became apparent that multiple departments could benefit from this project, mainly because so little GIS knowledge is necessary to navigate the web application.

The Agua Caliente Land Status is the most critical layer used by the tribal government staff. It is used on a daily basis to answer questions involving realty, economic development, historical preservation, and environmental protection. Geoprocessing and analysis are frequently used to determine how reservation parcels are affected by boundaries and restrictions placed by overlapping jurisdictions such as land use, endangered species critical habitats, zoning ordinances, and flood zones.

Numerous GIS products and services have resulted from this project, many of which have changed throughout the years with evolving technology. All planning and development team members have a current wall-mounted map hanging in their office and ArcGIS installed on their computers with access to the user database. They also have online access to the land status web map application that provides links to all related property documents. The tribal government leadership benefits from these products and services in numerous ways. They benefit financially by having an in-house GIS that can produce maps and exhibits that support tribal-approved plans without having to outsource the work. The Tribe also benefits from having desktop access to the most current reservation property information, which allows for more informed decision making by the planning and government staff. Current and accurate representation of the reservation properties is not only an asset to the Tribe, but also the Tribe's neighboring jurisdictions. The Tribe shares this data with the local cities and government agencies, including the BIA, and makes a PDF version of the land status map available on the Tribe's website for the public to access.

The next step for the project is to make this information available, through a web mapping application, to the individual tribal members. Currently, this information is only accessible to the tribal government, but the Tribe has created a membership website as a resource to tribal members. The goal is to eventually link this website to a mapping application where tribal members can easily retrieve detailed information about their own properties.

A GIS program can actually help a tribal government grow while retaining cohesiveness between departments. The Agua Caliente Tribe GIS staff imagines a day when GIS will provide complete connectivity of tribal land dealings across all departments. This would not only streamline workflows like realty transactions and eliminate redundant data, but would also provide the Tribe with real-time decision support in a changing world.

Use of GIS to model land acquisition scenarios

Brady Smith, GIS Planner; and Richard Colvin, Environmental Planner, Confederated Tribes of Siletz Indians
Siletz, Oregon

When the Confederated Tribes of Siletz Indians (Siletz Tribe) regained status as a federally recognized Tribe in 1977, it had no reservation. In the 1980s, through special legislation, the Tribe received approximately 3,630 acres of federal land (primarily timberland) for a reservation, with revenues from timber harvest funding, essential government services, and some very basic social services. Since 1977, purchasing land for housing, economic development, and government operations has remained a high priority, and the Tribe has acquired over 5,000 acres of land for housing, commercial development, and natural resource management.

Today, land acquisition is necessary to support the tribal treasury, provide adequate housing, increase capacity to deliver vital services to the membership, and foster tribal culture. Expansion of the tribal land base is vital to establish authority for asserting jurisdiction over resources, including land use, and to qualify for federal assistance to increase service capacity. However, not all land is suitable for tribal purposes. The Siletz Tribe saw a need to evaluate potential land acquisitions, based on foreseeable program demands, for the purpose of targeting acquisitions in order to meet specific needs of the tribal community.

The Tribe realized that a consistent methodology for evaluating potential land acquisitions would be valuable. In 2003, the Tribe received a grant from the Administration for Native Americans to develop a land acquisition plan that would describe a process by which potential properties could be evaluated. In June 2005, the final plan was approved by the Siletz Tribal Council. The approved plan sets forth procedures to follow when evaluating land for acquisition, including the types of maps, data, and land-use information that is to be collected. Since 2005, the plan has guided the Tribe in acquiring land based on the intended use. In addition, the plan serves as a tool that helps decision makers determine which properties will best achieve tribal needs.

As part of the planning process, a baseline assessment was conducted, where the Tribe evaluated the overall availability of land and the best properties to target for development within the Tribe's area of interest. The baseline assessment evaluated a variety of existing land-use types, including vacant, farm, and underdeveloped land. Various GIS layers of resource information were collected to determine the ease of development. The GIS layers included physical characteristics (slope, soil stability, hydric soils, floodplain), utility availability, and desirability (lot size and shape, market value, and proximity to other tribal property). The GIS program then developed a model that scored each parcel based on criteria established in the land acquisition plan.

The baseline assessment allowed the Tribe to identify properties that would be best to purchase for development. In addition to a land acquisition priority list, the Tribe is able to evaluate various criteria if a specific need for land is identified. The database also allows the user to develop acquisition scenarios based on need and to evaluate numerous properties at one time.

As with any assessment, there are limitations on its use. Since the assessment used mostly existing data from various different sources, some of the layers were developed at a scale not intended for analysis of small individual tax lots. This had the potential to make the results different than real world conditions. Thus, some amount of field verification was required. Depending on the layer, field verification could be as simple as a visual inspection to ensure that utilities are indeed there, or it could be more complicated like site-specific soil analysis to determine a site's suitability for building.

The use of multiple proximity analysis tools and database calculations used in the creation of the baseline assessment scoring criteria made it difficult to continually update the database when new information became available. The assessment of the property also removed the visual evaluation of the information, which allowed the user to interpret the information. For these reasons, the Tribe did not continue to update the assessment after it was initially created. Having the layers readily available to perform evaluations on individual properties was deemed more important than a table that scored each individual property.

To address the need for standardized evaluation procedures for potential land acquisitions, the Tribe used layers from the baseline assessment to develop an initial evaluation. Since the form is completed with existing property information, the initial evaluation can be done without visiting the property. If the evaluation suggests that a property be given further consideration, field visits may be appropriate.

Today, the data is available to tribal management and staff through a published map read by the ArcReader application or through a web map powered by ArcGIS Server. These read-only applications are available to staff, so they can access information about a potential land acquisition. The availability of these tools that allow non-GIS people to access the information has reduced the overall information requests pertaining to real estate received by the GIS staff.

Establishing legal boundaries of unsurveyed tribal trust lands in the Nooksack River Delta

Gerry Gabrisch, GIS Manager, Lummi Nation
Bellingham, Washington

The Lummi People are a Coast Salish People whose traditional territories include the San Juan Islands and the waters and coastal watersheds of Puget Sound from the Fraser River south to the environs of Seattle, Washington. The cultural and historical traditions of the Lummi People reflect their connection with and dependence on nature for their well-being and survival.

In 1855, the Lummi People signed the Point Elliot Treaty with the United States, which established the current boundaries of the Lummi Indian Reservation. The Lummi Reservation is located approximately eight miles northwest of Bellingham, Washington, and includes 12,500 acres of uplands and approximately 7,000 acres of tidelands. The Lummi Reservation comprises the Lummi Peninsula, the surrounding floodplains of the Nooksack and Lummi Rivers, a northwestern upland area, the Sandy Point Peninsula, Portage Island, and the surrounding tidelands.

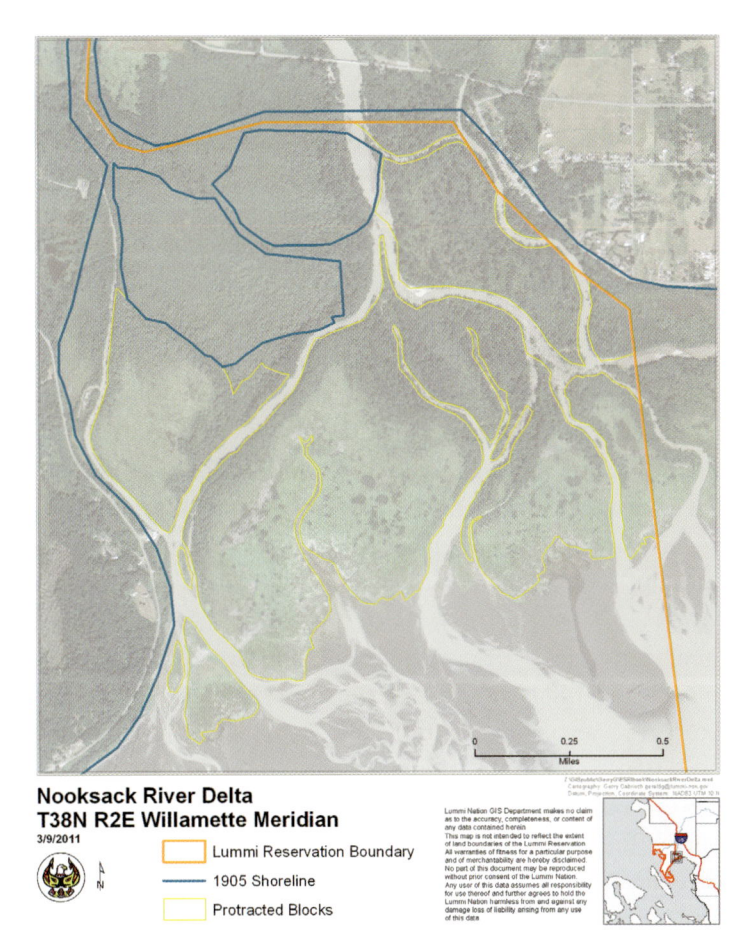

Chapter 4: Economic development on tribal lands | 59

Figure 3. Nooksack River Delta with the 1905 shoreline. Courtesy of Lummi Nation

Although the Lummi Nation started using a GIS in the mid-1990s, the Lummi Nation GIS Division was not formally established until 1999. The division's mission is to develop, document, and maintain spatial information for all Lummi Indian Business Council (LIBC—the elected governing body of the Lummi Nation) departments, and to promote the use of spatial information and a GIS where practicable to serve, meet, and protect the spiritual, economic, social, cultural, educational, physical, and environmental needs and values of the Lummi People. Current GIS staff includes one full-time GIS manager, one half-time GIS technician, and approximately twenty GIS users of varying skill levels.

One example of how the Lummi Nation uses a GIS involves the establishment of a legal boundary for the Lummi Reservation in an area of accreted tidelands at the mouth of the Nooksack River. The historical legal boundary of the Lummi Reservation was established by traditional surveys beginning in the mid-nineteenth century continuing into the mid-twentieth century. The eastern-most boundary of the Lummi Reservation followed what, at the time, was the left bank of the Nooksack River to the tribally-owned tidelands in Bellingham Bay. Since the adoption of those surveys, deposition and accumulation of sediments from the Nooksack River have added a substantial amount of upland acres to the Nooksack Delta, expanded the tribal tidelands within the delta, and altered the course of the Nooksack River.

The Lummi Nation is developing a wetland and habitat mitigation bank, and a portion of this mitigation bank is located in the Nooksack River Delta, including the newly accreted lands. Because these lands were deposited after the most recent surveys in 1937, a formal legal description and land ownership title did not exist for this area. Since a formal legal description and land ownership title is a required element for the Mitigation Banking Instrument, the BLM was contracted to help establish a legal description and determine the ownership of the new uplands. The work by the BLM allowed the continued development of the wetland mitigation bank by updating and accurately defining the boundaries of the Lummi Reservation.

Because of the prohibitive costs of a traditional land-based legal survey, the difficult terrain and associated access limitations, and the dynamic nature of the Nooksack River Delta, the BLM decided to use a legal description of the boundaries for the Nooksack Delta uplands developed by the Lummi Nation GIS Division. This legal description is based on a combination of high-resolution aerial photography (6-inch resolution) and light detection and ranging (lidar) surface elevation models. The legal description based on the remote sensing information defines the boundaries of the additional uplands area in the Nooksack Delta and facilitates the establishment of the Lummi Nation wetland and habitat mitigation bank.

To develop the legal description of the boundaries of the accreted lands, the Lummi Nation GIS staff created a metes-and-bounds generator using the tribal GIS and the Python programming language. The subroutine automated the calculation of the distance and azimuth of the Nooksack River Delta upland site boundary. Using lidar and high-resolution oblique and orthogonal aerial photographs, the Lummi GIS Division created a GIS layer of points that outlined the upland boundary of the Nooksack Delta. The latitude and longitude of the points were written to the attribute table of the GIS layer. Next, the data was reprojected into a Washington State Plane projection, and coordinate values in feet were calculated and added to the GIS layer. Once the data preprocessing was completed, a Python script was developed that enumerated the attribute table and captured the latitude, longitude, and state plane coordinates, and then calculated angles and distances from the sequentially ordered points. A combination of trigonometric functions was used to generate an azimuth from true north from point 1 to point 2. Next, the state plane coordinates were used in a simple distance formula calculation from point 1 to point 2. The resulting azimuth and distance from point 1 to point 2 were written to new columns in the GIS layer's attribute table. This process was repeated for all points until directions and distances were calculated for all points. Finally, the attribute table was opened in a spreadsheet and the text for the metes-and-bounds description was added. The resulting file was converted to a text document and submitted to the BLM for approval. The BLM reviewed and signed the final plat based on its legal research, ownership interpretation, and the provided legal description in the form of an amended protraction diagram in early 2010. The amended protraction diagram was filed in the Federal Register, formalizing the change of the Lummi Reservation boundary to include the certain lands accreted to the Nooksack River Delta, and the amended protraction diagram became formally adopted on May 3, 2010.

This project has resulted in a more accurate understanding of the tribal lands, which allows the Tribe to more effectively manage its resources. In addition to using GIS to better define the reservation boundaries, the Lummi Nation uses GIS to support land-use planning, forestry, fisheries, tidelands, restoration, utilities, hazard mitigation, economic development, and water resources management.

A new era of land-use planning

Wynette Arviso, JJ Clacs & Company, Navajo Nation
Fort Wingate, New Mexico

The Navajo Nation operates a sophisticated sovereign government responsible to approximately 198,000 residents on the largest federally recognized American Indian Nation in the United States. Spanning portions of Arizona, New Mexico, and Utah, the Nation covers roughly seventeen million acres. In addition to five administrative agencies, it is divided into twenty land-use districts and 110 Chapters. Each Navajo Chapter exercises local autonomy for services specific to its membership. As a sovereign Nation, the Navajo manage their own resources and infrastructure, which is a task that has become more complex over the years as the need for economic development competes with the need to preserve natural and cultural resources as well as traditional lifestyles.

Managing these issues requires a sophisticated understanding of how the various themes play out across the landscape. The Nation is constantly faced with challenging questions where the geography of the Nation plays a central role:

- Where are the grazing districts in relation to the mining operations and traditional cultural properties?
- How will new business opportunities be appropriate and consistent with current land uses?
- Will transportation and energy levels support development without new construction of roads and power lines, and in what way will people be impacted?

To create development that is appropriate for the community but broad enough to draw the desired long-term effect, questions like these must be answered by well-informed decision makers. In all cases, the answers to these questions are as unique as the community they address and the people who live there.

Like many Native American communities, the Navajo People traditionally managed land issues based on oral language. Traditionally, property boundaries and grazing districts were recognized and agreed upon verbally through the identification of landmarks, and these agreements were binding. Families alone were able to keep track of who had interests in the various tracts of land. Even after the United States established reservation boundaries in 1884, Navajo People continued to keep verbal tabs on their holdings. However, in order to implement greatly needed economic development across the Nation, new methods were needed to adhere to outside funding requirements of the Local Governance Act (LGA) of 1998. This act gave Navajo Chapters the ability to take greater responsibility for land-use development and administration. As amended in 2001, the law requires that communities that receive funding must develop community-based land-use plans. Because oral tradition was the dominant land management tool for so long, actually writing down land-use plans was an entirely new experience for the Chapter leadership and members. Faced with following a public-driven process and identifying land use based on the community's vision, goals, and objectives, tribal leaders undertook the arduous challenge of explaining this new process to their membership. Although the process was new, which in itself presented its own set of obstacles, the open communication was critical and has since become one of the most important factors responsible for the Navajo's success in economic, community, and land-use planning.

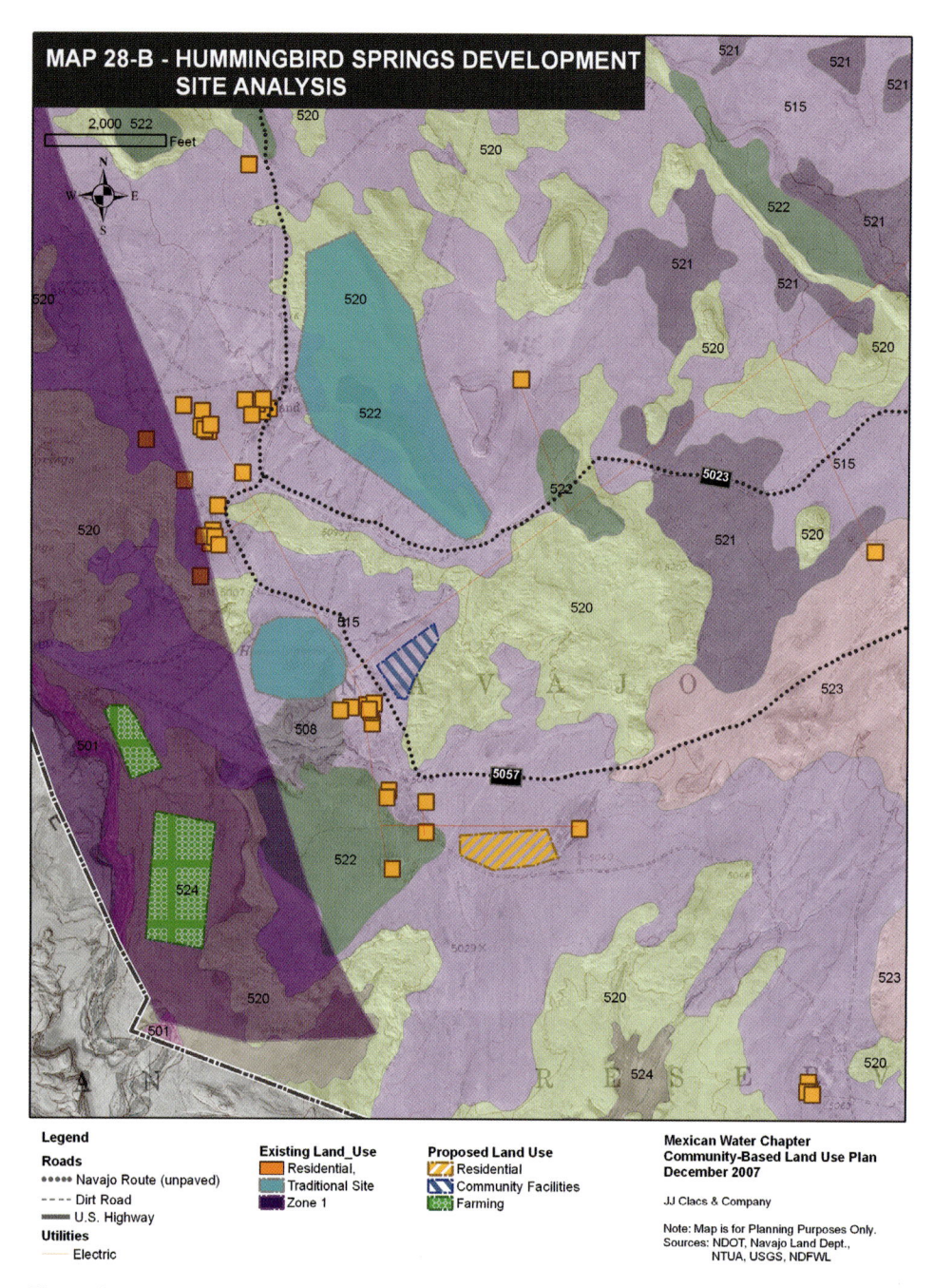

Figure 4. Overview of the Hummingbird Springs development. Courtesy of Wynette Arviso, JJ Clacs & Company

Three aspects of communication have proved integral to project success. First, GIS technology provided a much needed visual strategy that proved compatible with Native oral traditions, helping bridge the gap between the traditional and modern worlds. It also allowed Chapters to incorporate their traditional customs while articulating the community's overarching goals, objectives, and strategies to guide and coordinate land uses. Second, translating the process into the Navajo language significantly increased participation and allowed tribal Elders and Native-only speakers to participate. This participation by tribal Elders who only spoke Navajo was an important endorsement of the process to the rest of the community. Lastly, a Native-owned-and-operated consulting team, JJ Clacs & Company, facilitated bringing community leaders and members under one roof, helping the Chapters navigate this new territory and successfully plan their own futures. While each of these components lent critical support to the land-use planning process, the focus herein is on the role of GIS and how the introduction and development of geospatial technology on the Navajo Nation launched a new era in land-use planning.

It has been nearly a decade since JJ Clacs & Company introduced GIS technology to aid in the development of land-use plans for some Navajo Chapters, including Forest Lake, Beclabito, Chinle, and Shiprock. For these Chapters and many more, GIS has become essential in the community outreach aspect of planning because through GIS products, residents who typically reside year-round in rural areas and who may even rotate between traditional summer and winter homes were able to view their grazing districts, land ownership, agency jurisdiction, traditional areas, modern facilities, and community centers in a single comprehensive view for the first time.

In one example, residents of the Two Grey Hills Chapter, who summer their sheep in the Chuska Mountains and winter them in the valley, now use GIS to illustrate their grazing practices in order to guide future management decisions. Illustrating community land-use decisions such as the location and direction of houses relative to hogans (traditional structures), roads, water lines, sewage, and electricity complements decision making based on oral traditions because GIS allows the membership to view the entire community at once. This view is invaluable when considering how new developments will affect the traditional use of the tribal lands. GIS offers a visual tool for helping people understand land-use planning as it is commonly applied in nonnative settings.

GIS technology has also facilitated the education and outreach for a large and complex community-based land-use plan with twelve potential development sites for the Mexican Water Chapter. The initial step in developing the future land-use component was to identify the values of the community members and what was important to them. Such envisioning and planning was particularly important in considering the potential for future growth and development. To this end, community members gave their input through public meetings and work sessions. Subsequent public hearings were held for the purpose of identifying a vision for the community that would reflect the desired end uses for the various land areas.

Speaking in Navajo at planning meetings like those at Mexican Water, the consulting team explained how the process works and how the maps are generated. They used ArcGIS Desktop software to illustrate Chapter boundaries previously identified through oral traditions, identify land status, and map natural and cultural resources and infrastructure, such as water and power facilities and transportation systems. Initially, existing data was obtained from the community as well as the BLM, Navajo Department of Transportation, the Natural Resources Conservation Service's Soil Survey Geographic (SSURGO) database, New

Mexico Resource GIS Program, and Arizona Regional Image Archive, among others. Much of this information was then overlaid on either oversized topographic maps or aerial photography and presented at public meetings for the community's input.

After becoming comfortable with the mapping data, leaders helped identify the Chapter boundaries for the planning area, since there are no legally surveyed boundaries for these areas. The community members then pointed to various locations, discussed what had transpired at those sites, determined if any areas were sacred, and what areas could be considered for other uses. This data was combined with information from other sources such as roads, utilities, and water wells to determine the appropriate pathway forward for land-use purposes. Participants were also encouraged to hand draw their planning suggestions on oversized wall maps. When an official decision was made regarding the planning unit's final boundaries and components, the new information was digitized in ArcGIS Desktop.

Typically, planning projects are funded by the Chapter undertaking the project. The principals of the consulting firm collect and assemble the GIS data, and with the help of other Native speakers in the community, present the maps to participants. They then digitize the results. Ultimately, the data is transferred to the local Chapter, and hard copies of all maps and planning documents are provided as a basis for local management and updates to the data.

Mexican Water's future land-use plan is designed to inspire ideas that provide a broad, yet clear picture of the community as its members, leaders, and the general public envision it to be. The community-based land-use plan is also the community's guide for managing growth in the location, type, scale, and density of future land development. The maps used in the plan indicate the intended predominate future function, density, and characteristic use of the land. In this case, they do not reflect the intended zoning of individual areas but rather generalized desired future land uses. In some cases, the maps are also used to indicate an overall mix of population and business densities.

The use of GIS mapping for long-term planning on the Navajo Nation is unprecedented, and the benefits, some expected and some not, extend much further. The most notable benefit of GIS on the Navajo Nation is a published land-use plan that has tremendously enhanced the way Chapter officials and community members make short- and long-term land-use decisions. Prior, the Navajo People conducted land-use planning based solely on oral traditions that have been and still are passed down from generation to generation. To this day, Chapter boundaries are not physically recorded in the same way jurisdictional boundaries are documented off the reservations.

The visual nature of GIS also helps bridge language barriers at public meetings where some community land-use planning committee members, as well as many public participants, are Native-language-only speakers. The maps developed in GIS provide the needed tools to encourage these individuals to participate throughout the projects and eventually present the chosen land-use plan and GIS information in Navajo to other Native-language speakers. Further, GIS captures and graphically displays information that Elders and Native-language-only speakers convey during the planning process. In other cases, GIS-based land-use planning enlightens Chapter members when they are able to see the actual size and shape of a particular landmark, grazing unit, or geographic feature. Thus, presentation of oversized land maps easily communicates information for Native-language-only speakers and encourages greater participation in land-use planning.

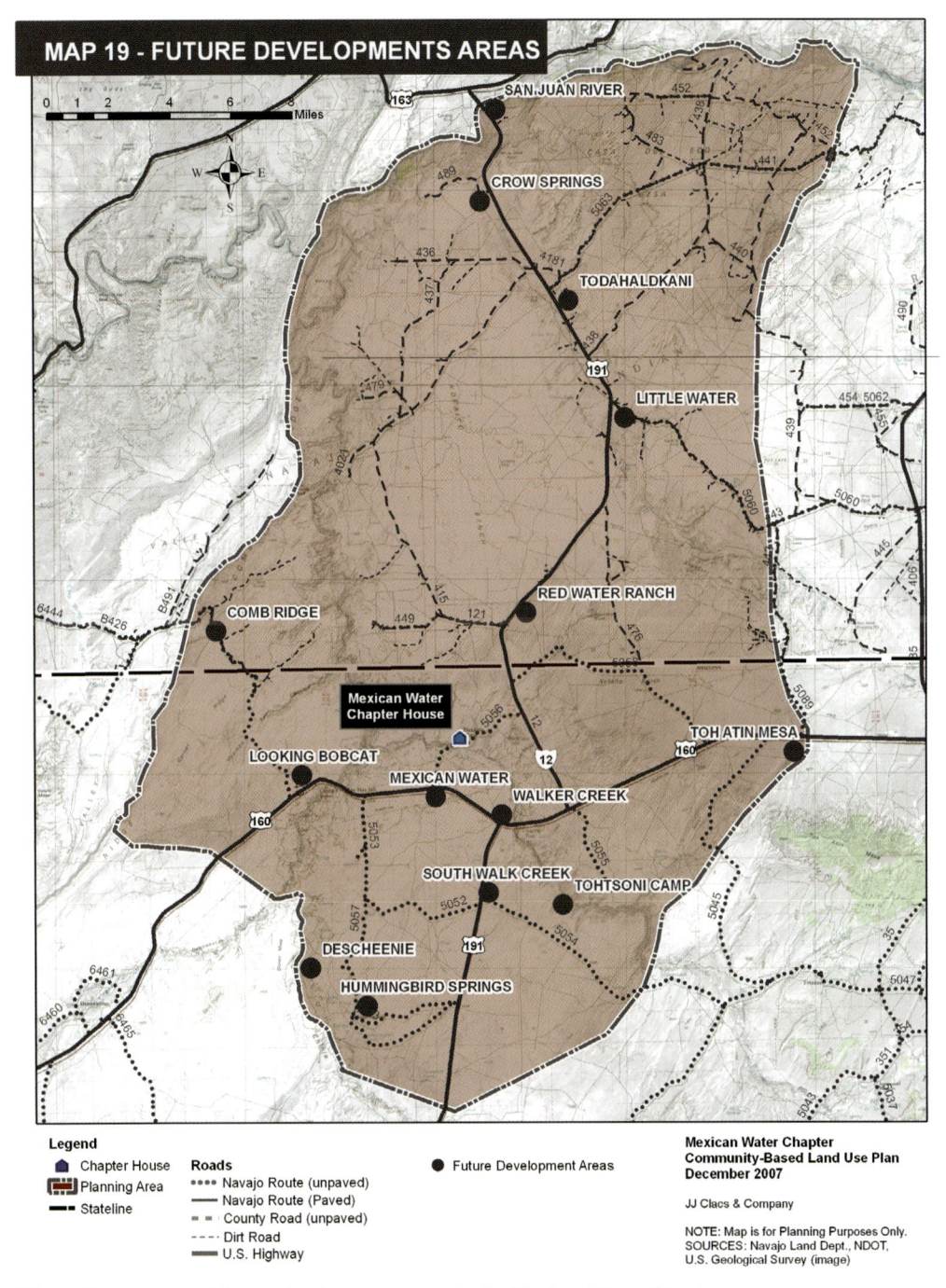

Chapter 4: Economic development on tribal lands | 65

Figure 5. Overview of future development areas in the Mexican Water Chapter. Courtesy of Wynette Arviso, JJ Clacs & Company

GIS technology enables the Chapters to develop effective land-use plans needed to receive the necessary certification under the LGA and become eligible for various federal and tribal funding sources. Ultimately, it allows Chapters to be more self-sufficient, sovereign, and economically stable for future generations. These examples demonstrate that the benefits of incorporating Native and nonnative observations and research through GIS are far-reaching. The overwhelming success and widespread acceptance of GIS technology on the Navajo Nation can be attributed to not only the Chapters' perseverance and the consultant's effective project management but also the surprisingly compatible nature of GIS as a tool in Native American resource planning. Indeed, Native Americans are visually oriented, and GIS provides the Navajo Nation's Chapters with the opportunity for them to sustain their own lands and respond more effectively to their own needs while becoming more competitive in both Native and nonnative venues.

Creating healthy communities

The geographic dimension encompasses many scales. On a personal scale, where we live, what we eat, and our behavior all affect our health. On a local or regional scale, public health officials and others must decide where to locate a new health facility, or how to efficiently support health services for a certain area or certain segment of the population. On a continental or global scale, the monitoring of the origin and spread of epidemics must also account for geographical factors. Because of these and many other health-related issues, GIS has become an essential tool and framework for understanding problems and issues and in designing effective health policies and interventions for tribal communities. Furthermore, the ability of GIS to deal with data that is incomplete, or that changes rapidly, becomes especially important in its application to health, because health data is typically incomplete and can change on a daily basis. Being able to map and analyze the spatial pattern of where hospital and clinics are located relative to the age and other characteristics of the population, where specific kinds of diseases are most prevalent, or even where sources of pollutants exist, are all tasks that a GIS can help decision makers accomplish. Indeed, one of the earliest documented examples of the map overlay process that GIS later made use of was in the field of public health. In 1854, Dr. John Snow plotted cholera deaths to water pump locations in London, showing that the disease was clearly linked to a waterborne pathogen. Today, tribal governments and researchers are taking advantage of GIS in a wide variety of ways to better understand and improve tribal health needs.

The stories in this chapter show how tribal governments and researchers use GIS to enable them to make the best decisions possible regarding one of the most precious of resources—human health.

Jen Olson, epidemiologist, describes how the South Puget Intertribal Planning Agency uses GIS to better understand the incidence and pattern of cancer among the Chehalis, Nisqually, Squaxin Island, Skokomish, and Shoalwater Bay Tribes of Southwestern Washington. They use this data to design programs to more effectively meet the needs of the more than 13,000 Native Americans in this region. On the Fort Peck Reservation in Montana, Allyson Kelley, Ken Hull, Gary James Melbourne, and Eric Wood use GIS to identify the environmental determinants of childhood and adult asthma. On the Standing Rock Reservation in North and South Dakota, Dennis Painte describes how he is compiling a GIS database to understand the adverse health effects of lead paint poisoning.

By embracing GIS technology, a Tribe can assess the health needs of its members but also incorporate local traditional knowledge and communication networks. Traditional knowledge is critical in the development of personal health histories, and GIS can help a Tribe connect Elders' stories with current situations in an effort to improve the health of everyone. The stories show that gathering data based on a geographic location can be a long process, involving surveying the population and often requiring visiting individuals and families. This can take many years and many miles of travel. However, being able to "geographically enable" health data is so valuable that dedicated researchers and practitioners have put in countless hours to locate and map individual data.

Human health is a sensitive issue, and these authors demonstrate how GIS can be used to handle the sensitive nature of health data while also using it to help people in need. Human health is also a complex issue, resulting from the environment, culture, behavior, family history, and location. GIS provides the framework that enables people to see across these complex interrelationships and dependencies. GIS enables the integration of diverse layers of information into a common operational view where patterns, linkages, and trends can be given a context. GIS can also be used as a tool to present a compelling case to funding agencies that the problems and issues identified require additional funding to successfully carry out the recommendations suggested by the project managers.

Each story shows how GIS can be applied to health challenges. Through the GIS work that they are doing, human lives are valued, and the health of the generations to come can be improved.

Planning for healthy communities with GIS at the South Puget Intertribal Planning Agency

Jen Olson, Epidemiologist, South Puget Intertribal Planning Agency

Tribal history

Tribes along coastal Washington have a long and rich history. Like many Natives of the northwest coast, these Tribes relied on fishing from local rivers, shellfish from the coast, camas root, and berries for food. They hunted, raised horses, and built plank longhouses to protect themselves from the elements.

Chapter 5: Creating healthy communities | 69

Figure 1. The lands of the coastal Washington Tribes are rich in natural resources and history.
Photo by Joseph Kerski

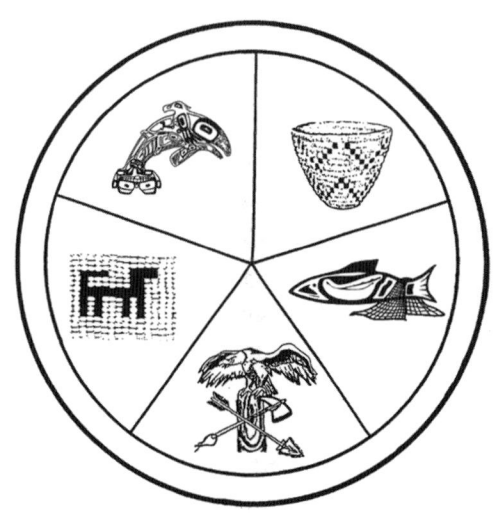

Figure 2. Logo for the Five Tribes of Southwestern Washington. Courtesy of South Puget Intertribal Planning Agency

In the Coast Salish language, map is *scqiulex^w*, a writing or marking. As a prefix, it is *sc*—that has been done, made, or prepared. The word *uleö* as a suffix means land (Pete 2011).

The South Puget Intertribal Planning Agency

The South Puget Intertribal Planning Agency (SPIPA) is a tribally chartered nonprofit organization serving the Chehalis, Nisqually, Squaxin Island, Skokomish, and Shoalwater Bay Tribes of Southwestern Washington. SPIPA has been planning, implementing, and operating programs with the Tribes since the signing of its founding compact in 1976. SPIPA has a close working relationship with the Five Tribes and helps to build capacity for health and social service programs supervision at the local level.

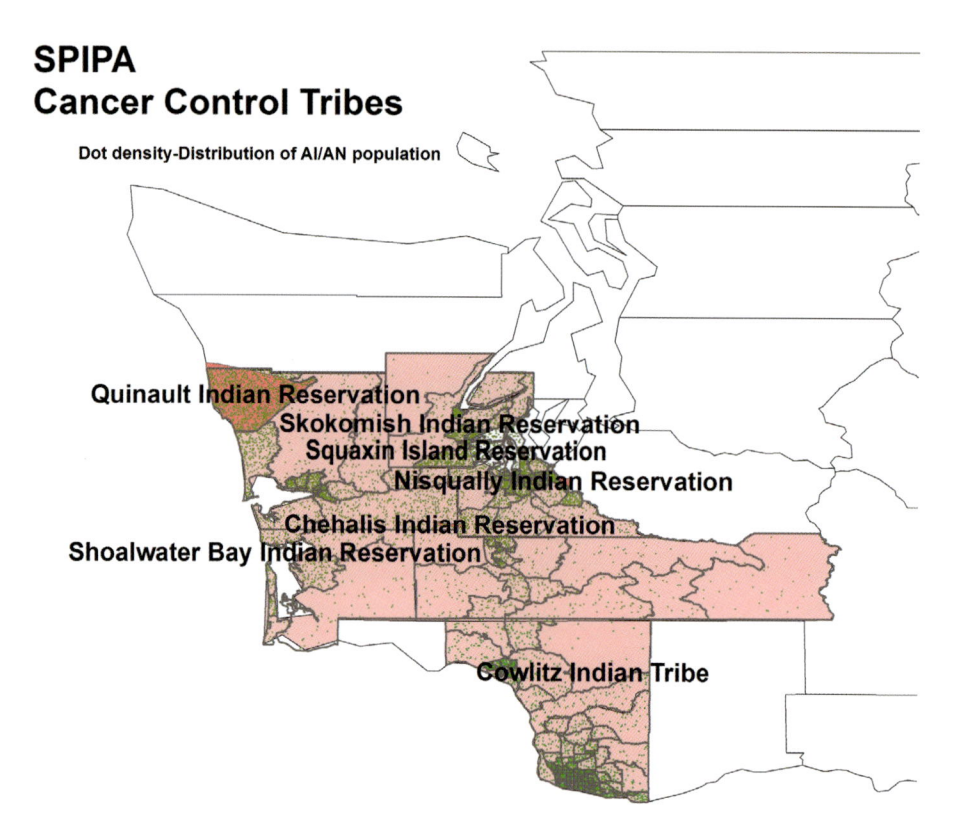

Figure 3. Service area for SPIPA showing a dot-density distribution of American Indian/Alaska Native population in the counties it services. Courtesy of South Puget Intertribal Planning Agency

SPIPA-managed health, social service, education, and training programs provide services to more than 13,000 Native Americans throughout the southwestern Washington counties of Pacific, Lewis, Mason, Thurston, Pierce, Grays Harbor, Cowlitz, and Kitsap. One of the many programs at SPIPA is cancer control. Through the combination of three Centers for Disease Control (CDC) funded programs in partnership with tribal councils and tribal health clinics, SPIPA is able to provide comprehensive cancer control services to this large tribal population.

SPIPA began cancer control services in 1994 with a women's breast and cervical cancer screening program called the Native Women's Wellness Program (NWWP). Over 2,775 women are currently enrolled in the NWWP. SPIPA subsequently received funding for cancer control planning, implementation, and resources to include colon cancer screening.

SPIPA began using GIS in 2004 and currently uses GIS for its cancer control programs. Geographic location is one of the most common variables considered when collecting and presenting health service data, disease rates, and risk and behavior data to the Five Tribes. GIS has allowed the agency to conduct data analysis by location, link to other data such as census data and vital statistics data, and effectively present conclusions to the Five Tribes.

As the Consortium of Five Tribes, SPIPA provides geographic analysis and presents data as important components of its services. Each Tribe has its own identity and defined community. Using GIS to analyze and present data that coincides with tribal communities assists in defining needs, managing existing resources, and graphically communicating issues with tribal councils and communities as well as outside partners and granting agencies. GIS software was provided through the Esri Conservation Program.

SPIPA currently uses GIS for a number of health projects. The SPIPA Comprehensive Cancer Control Program began using GIS during the planning phase of cancer control. The program initially partnered with the Washington Department of Health to generate maps displaying the top three cancers for areas with a higher than average percentage of American Indian/Alaska Native (AI/AN) population. The cancer control program serves not only tribal members living on the reservation, but extended "community members." For program planning purposes, the definition of community member includes the entire population living on the reservation as well as the populations living in census tracts surrounding the reservations with a higher than expected rate of AI/AN population, according to the 2000 census. The resulting map of leading cancers, in rank order per identified tribal area, helped the communities prioritize target areas for cancer prevention and control.

GIS design

GIS has helped the cancer control program illustrate the burden of disease for tribal service areas. The program has generated maps from state, county, and ZIP Code level health statistics. These maps have been used in presentations, applications for funding, and in reports to the Tribes alongside or instead of graphs or data tables. Elders in the communities have preferred to see cancer statistics presented on maps, with their reservation clearly outlined.

The cancer control programs have linked and mapped cancer incidence data from the Washington State Cancer Registry with census track data for the Five Tribes. This information is updated periodically, and new maps are generated and shared with the community.

Lung cancer is the leading cause of cancer-related death among five participating Tribes. Commercial tobacco use, specifically cigarette smoking, is the leading cause of lung cancer. The maps showing lung cancer as the leading cause of cancer death have been useful to convey the message that one key way to prevent cancer is to prevent or help stop the use of commercial tobacco by adults and youth. This

Figure 4. A map of leading cancers for AI/AN men in Washington was generated and presented during the annual Native Men's Wellness Workshop to illustrate the impact and comparison of lung, prostate, colorectal, and liver cancer of the AI/AN population in Washington. Courtesy of South Puget Intertribal Planning Agency

cancer information, geographically referenced to the location of the tribal communities, has served as a powerful message in helping the Tribes understand the impact of cigarette smoking and secondhand smoke in each community.

GIS implementation

Primary GIS personnel

The CDC funds breast, cervical, colorectal cancer control for the SPIPA Tribes. A half-time staff epidemiologist for these programs is the sole person trained in and using GIS at the agency. The cancer programs work with the SPIPA epidemiologist to monitor all screening, cancer incidence, and mortality within the combined service area. The cancer control epidemiologist is able to use GIS for data analysis and mapping. As the wellness programs expand, the epidemiologist hopes to train other staff members, and ideally tribal youth, in using GIS for health and wellness.

SPIPA has been collecting breast and cervical cancer screening rates, clinical data, and cost data since 1995. The Cancer Screening and Tracking System (CaST), a CDC-sponsored electronic patient data system is used for this purpose. Two of the seven tribal clinics in this project have initiated the

Electronic Health Record (EHR) system recommended by the Indian Health Service. As more of the clinics implement the Electronic Health Record, SPIPA cancer programs will help coordinate the tracking of data elements with the goal of tracking all elements at the clinic level. This will help avoid duplication of data entry, enhance case management, and support local ownership of data.

ArcGIS is used to create maps used by the program. In order to use data from existing SPIPA databases and sources, the data tables are converted to database formats and "joined" to existing data tables that include geographic information, such as the data tables found in geocoded US Census layers. The program has used a variety of symbology, such as dot-density, to illustrate health data such as cancer incidence and clinic service population. In the map shown in the photo, cancer "counts" of actual cancer cases were illustrated using graduated circles. The size of the circles was customized so that each circle size represented a constant number of cancer cases. Manually controlling the size of the circles was needed to present the comparison of cancer via overlapping representative circles.

Workflows

In addition to data collected by SPIPA, wellness data, including cancer incidence and mortality, is publically available at the county level by several demographic factors, including race and sex. After US Census data, cancer incidence and mortality by county and by race has been the most-used data by the SPIPA programs. County level cancer data, by race, is available in many states. SPIPA and the Washington Department of Health finalized a data-sharing agreement in the fall of 2009 to allow SPIPA access to county-level vital statistics. Ideally, SPIPA and the partnering Tribes would have access to cancer data and vital statistic data at the ZIP Code or census tract level. Accessing health-related data held by other agencies, including state and federal government, is not an easy task for Tribes and tribal organizations. This is due in part to protection of patient confidentiality and possibly to limited access to state agency geocoded or health data at the Zip Code level.

Challenges

A few of the challenges for the program are due to the large geographic area of the SPIPA service population. As the service area of the program becomes better defined, GIS will support more efficient allocation of resources and the tracking of change over time. Currently, in many cases the program uses county-level data for the AI/AN population. This county-level data includes AI/ANs from neighboring Tribes that are not represented in the service-area population, and nonnatives served by the SPIPA Tribes (from social ties) that are not included when the AI/AN population is examined. Another challenge to gathering data on the service population is that many tribal members may not live on the reservation, but nearby.

Ideally, SPIPA would be able to define to the census tract level all people in a SPIPA community. This would include all AI/AN, but not necessarily enrolled in that Tribe, as well as nonenrolled community members, those living on the reservation, as well as those living close to the reservation and still receiving services.

Results and benefits

The final outcome of the project has been a clearer understanding of the cancer incidence for the populations served by SPIPA and the Five Tribes. A tangible result of this data analysis, presentation, and discussions with the Tribes is a ten-year cancer control plan that has been successfully implemented for our Five Tribes.

Next steps

In the future, SPIPA plans to further leverage GIS technology in a variety of ways. One will be the use of GIS to map the results of a tribal behavior and risk factor survey. Because of the success and acceptance of GIS used for health data, the survey design included geographic data elements such as address and ZIP Code to enhance GIS analysis.

The SPIPA cancer programs will also be using GIS to analyze the location of those receiving cancer screening services in relation to the location of diagnostic services. This information will be useful for program planning to identify areas with adequate services within driving distance as well as areas in need of additional diagnostic services. The ArcGIS Spatial Analyst extension will be used to compare buffer zones of "likelihood to access services" in order to determine if distance from service is related to use of services.

The challenges that the cancer programs seek to use GIS to solve include analysis of cancer incidence, staging of cancer, and mortality rates in relation to distance from the tribal clinics, hospitals, and oncology centers. Nationally, American Indians and Alaska Natives are diagnosed at the latest stage more often when compared to other racial groups. Late-stage diagnosis is harder to treat and may require more intensive treatment than cancers caught at earlier stages. Access to screening and diagnostic and cancer treatment services are crucial to finding and treating cancer successfully. The SPIPA Comprehensive Cancer Control Program would like to determine if distance to (1) screening services, (2) diagnostic services, and (3) treatment services are related to early detection of cancers and completeness of treatment of cancers.

Each of the five SPIPA Tribes and tribal clinics collect varying degrees of health and wellness data tied to geography, such as a street address. Certain patient data elements, including address and diagnostic codes, are reported to the Indian Health Service but are not generally geocoded and made available to Tribes or tribal organizations for public health purposes. The next task for SPIPA health-related GIS projects will be to collect more accurate data, possibly using GPS when appropriate, such as for house-to-house surveys.

The SPIPA Board of Directors and agency planners have supported the use of GIS for health planning and envision using GIS for additional planning purposes. SPIPA fully utilizes census, tribal, and social services data for grant writing and program planning and evaluation. Future uses for GIS could include assisting with presenting data collected in other health service programs and social service programs. The agency has numerous wellness and social service databases that house geographic data. The potential exists to use GIS to conduct analysis and map additional services and remaining needs.

How GIS could assist Tribes

In order to have the greatest ability to study and monitor health trends within the participating Tribes, it would be ideal to have health data accessible and geocoded to the street or household for all tribal and community members. Much of the data on health and wellness is captured in a variety of databases or only in paper-based medical charts, some owned by tribal communities and some owned by outside sources such as research institutions and government agencies. Ideally, all Five Tribes will have the capacity, staffing, and community support to map the health data that will be collected electronically within the next decade. As the tribal health clinics expand to include electronic health records, Tribes and tribal organizations will not be dependent on state or national health data on their People. Once the data exists at the tribal level, the next step will be to use the data through programs such as GIS. The SPIPA Tribes, as well as many Tribes in the Northwest, have been using GIS for natural resource planning for a number of years. The SPIPA cancer control programs are the first to apply GIS for health data analysis and presentation in our area. We would like to encourage Tribes to think about using GIS for the data collected through their health and social service programs. Individual patient data (even if numbers are small) can be kept confidential through symbology, spatial analysis, and other tools, and can help define strengths as well as gaps in service.

Ideally, Tribes will not have to depend on outside agencies to report health data back to each Tribe. They will not have to depend on researchers or state health departments for GIS services for health and wellness. Each Tribe will be able to collect, geocode, and map data in order to communicate, assess, and address the health needs of their community members. This program has demonstrated the value of taking a geographic view of tribal health across a number of communities. With improved data and capacity within each tribal community, more can be done to ensure the health and wellness of tribal members.

References

Pete, Tachini. 2011. *Medicine for the Salish Language: English to Salish Translation Dictionary, Second Edition.* Pablo, Montana: Salish Kootenai College Press.

Analyzing asthma on Fort Peck lands in Montana with GIS

Allyson Kelley, Master in Public Health, Certified Health Education Specialist
Ken Hull, Registered Sanitarian, Fort Peck Environmental Health Officer
Gary James Melbourne, Fort Peck Tribal Health Director
Eric Wood, PhD, USGS Center for Earth Resources Observation and Science (EROS)

Where

Fort Peck is located in northeastern Montana, home to 10,321 people. This population includes 6,116 American Indians, of which 1,107 are Assiniboine alone, 3,406 Sioux alone, and 781 are of Assiniboine and Sioux ancestry (US Census Bureau 2000a). Fort Peck encompasses 3,289 square miles of rural lands (US Census Bureau 2000b). Poplar, Montana, a town of approximately 900 residents, is located at 48.11 North latitude, 105.2 West longitude at an elevation of 1,965 feet above sea level.

Since 1995, the Fort Peck Tribes Environmental Health Division has used GIS, allowing public health officials to visually and electronically record health information. In June 2007, the Fort Peck Tribal Health Department partnered with Rocky Mountain College, Fort Peck Community College, Indian Health Service, and the US Geological Survey (USGS) to conduct a pilot project to determine the spatial distribution of asthma on the Fort Peck Indian Reservation. Through this multi-organization partnership, asthma, a serious public health problem, is better understood.

Asthma is the most common chronic disease of childhood, occurring in approximately 54 of 1,000 children in the United States (Centers for Disease Control 2004). Increasing evidence suggests marked disparities among the prevalence of this disease among particular ethnic and racial groups (Mannino et al. 1998, Claudio et al. 1999). Understanding population and regional variances in the prevalence of asthma is important for optimal design of local interventional strategies as well as for elucidating important epidemiological insights into the disease.

Also of concern is the growing evidence linking asthma to environmental factors. A number of environmental agents such as environmental tobacco smoke, dust mites, pet dander, mold, and rodent feces have been implicated in the exacerbation of asthma. Control of the environment can significantly impact the expression and progression of the disease in people with asthma (Agency for Toxic Substance Disease Registry 2002).

Why

People in the community talked about asthma for over a decade, and it seemed like everyone knew someone with the disease. Prompted by these concerns, partners engaged in a retrospective review of asthma in the town of Poplar on the Fort Peck Reservation using GIS. The intent of this effort was to determine the spatial distribution of asthma in the community. The etiology of high rates of asthma in the community were not known; however, contributing risk factors such as particulate matter, poverty, housing stock, wind speed and direction, and close proximity to agricultural areas were thought to correlate with high-density asthma cases. A secondary question, which resulted from work in the community, questioned the impact of burned homes from arson on asthma prevalence (see this book's community mapping section in chapter 7 on higher education for more information).

The use of GIS was aligned with public health objectives through supporting the following:
- The assessment of the prevalence of asthma within the Fort Peck pilot study user population using ArcGIS
- The creation of a set of GIS data to show the spatial distribution of asthma prevalence for public health officials

Data from the Indian Health Service and the Fort Peck Tribal Health Department was used to plot the primary residence of asthma patients living in Poplar, Montana, during a specified time period. Ambient air quality threats, as a result of burned homes, were plotted at the request of the community using ArcGIS Desktop. The patient data for the Fort Peck Service Unit was obtained from 1988 to 2006. This data was reviewed by the project lead, confidential information was removed, and data was then compiled and mapped by the USGS.

In order to identify pediatric asthma patients, existing data was reviewed for patients aged zero to twenty years who have asthma (ICD -9-CM, code 493) listed as a diagnosis during their lifetime in the chart and at least one clinic visit during the calendar year in question (2005). The population-specific data was supported by two sources of data: (1) US Census Bureau block-level data for the estimated population density of participating reservation inhabitants under twenty-one years of age, and (2) the active user population of the service unit user population. Using the geocoding functionality of ArcGIS Desktop along with GPS locations, project leaders plotted waypoints of physical addresses of asthma cases in Poplar, Montana, on the map.

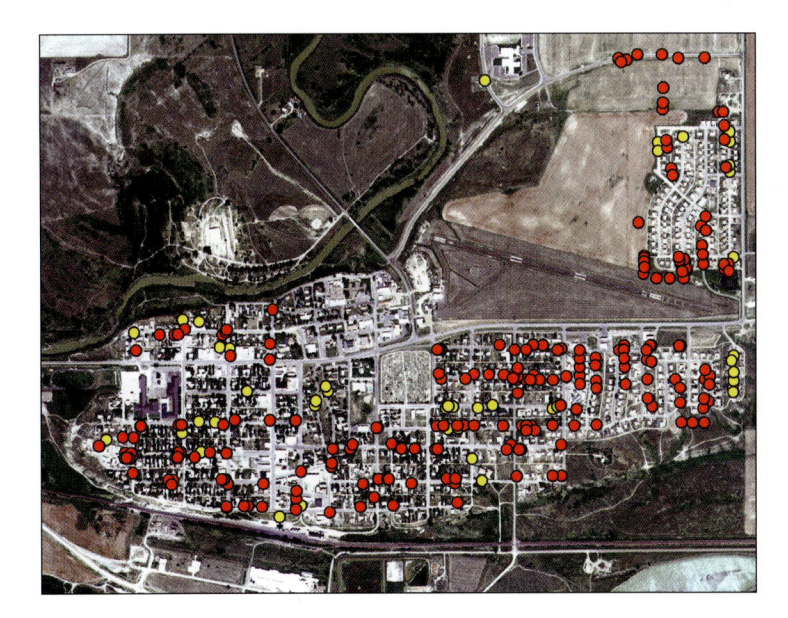

Figure 5. The location of asthma cases provided a baseline dataset for the study. Houses in yellow were obtained through field collection with GPS, and houses in red were obtained through address geocoding with ArcGIS. Courtesy of Allyson Kelley

Discussion

GIS was instrumental in helping stakeholders to understand the distribution of asthma in Poplar, Montana. This community was selected due to the relatively large population size, previous reports of high asthma prevalence, and staffing availability. Data from existing patient records was used to locate physical addresses for patients who were diagnosed with asthma for the 2005 calendar year. We then used GIS to manipulate the asthma data with census block density and ZIP Code data. The data displayed on the map reflects the initial results of the study. A welcome but unintended consequence of the project was the community's desire to become involved in GIS. Several community members wanted to know if a relationship existed between the number of burned homes and asthma cases. When homes burn due to arson or other factors, they emit particulate matter and other ambient air quality contaminants, which may adversely impact patients with asthma. Through a separate process, community members collected data for thirty-seven homes in the town of Poplar that had been burned due to arson. To answer the community's question about arson and asthma, a map was developed illustrating the asthma cases relative to the burned buildings.

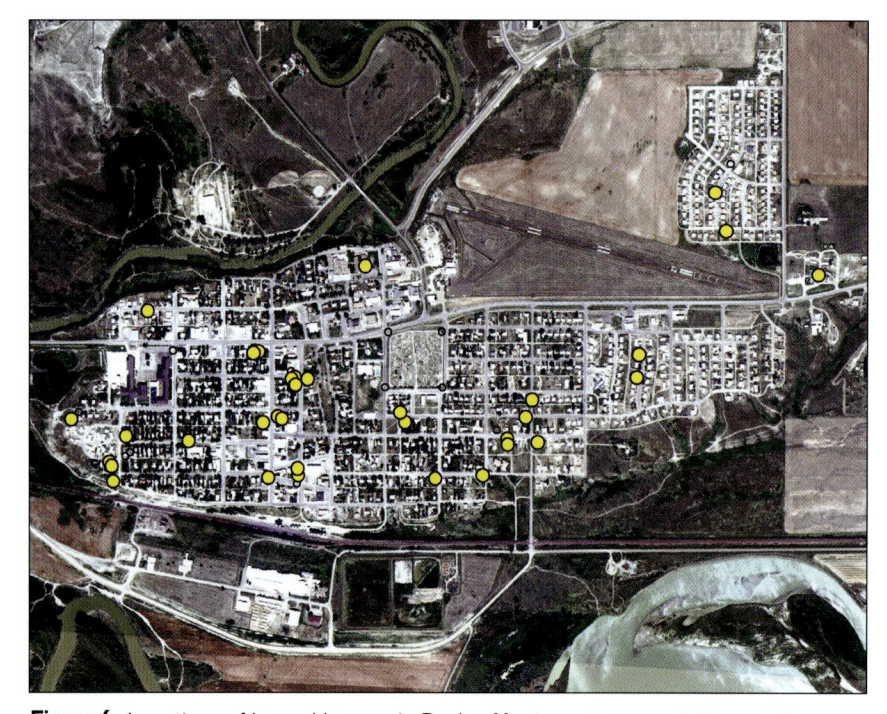

Figure 6. Locations of burned houses in Poplar, Montana. Courtesy of Allyson Kelley

GIS design

Several tribal government departments and other organizations within the Fort Peck Tribe are using GIS, including roads, housing, environmental, tribal health, minerals, Fort Peck Community College, and historic preservation offices. This project worked solely with the Fort Peck Environmental Health Department.

One challenge that faced the project team was that the majority of community members use post office boxes instead of physical street addresses. The reasons for doing this include the lack of a physical street address, because the house lacks a numbered address with a street name, moving frequently, or because of confidentiality. Without the physical addresses of asthma cases, it is nearly impossible to develop a map to relate environmental factors and housing location. Larger areas with multiple ZIP Codes often use the post office box ZIP Code to develop maps; however, the ZIP Code for Poplar, Montana, the town of interest, has only one ZIP Code for post office boxes.

Project partners deliberated on the most efficient method of collecting physical addresses for asthma cases. In the end, it was word-of-mouth and interviewing individuals that eventually provided the physical location of asthma patients for this project. The rural location and small population proved challenging throughout the project. As a result of geocoding in ArcGIS and supplementary data collected using a handheld Garmin GPS unit, the physical location of each residence was collected in the reservation's largest town, Poplar, Montana. The initial review included all towns and patients served by the Fort Peck Tribal Health Department and the Indian Health Service. We focused on the town of Poplar to map asthma cases because it provided the greatest number of physical addresses to correspond with reported asthma cases. Public health authorities were interested in Poplar because of the prevalence of disease found in this community and how disease may be influenced by housing, wind direction, agriculture operations, housing density, and poverty levels. GIS allowed the removal of confidential information so that maps presented to the community included only aggregate data.

Another challenge is that there is no definitive laboratory test for asthma; it is difficult to determine if all cases reported through the Resource and Patient Management System data query were in fact asthma or confused with another obstructive pulmonary disease. Additionally, the patient population fluctuates, and it is possible that the 2005 population estimate used in the denominator for calculating prevalence did not capture the total number of users in the Indian Health Service service population. Another limitation is the small numbers used to demonstrate asthma density in Poplar, Montana. The limitations regarding the use of GIS were (1) the small population, (2) the lack of physical addresses for many patients listed, and (3) the variance in case density is demonstrated, but the environmental exposures are not clearly documented in the map.

Ultimately, the product and the relationship between all partners grew as a result of this project. The chronic disease, asthma, which prevails in the community, is now a little better understood. The primary GIS personnel included one Indian Health Service intern, the tribal environmental health officer, and the USGS GIS representative. Various individuals from the Fort Peck Tribal Health Department staff assisted with data collection and analysis.

This project was made possible by the contributions of various partners and individuals. A Montana Indian Country Care Grant from the Environmental Protection Agency (EPA) supported part of the salary and materials for program staff at Rocky Mountain College. The Fort Peck Tribal Health Department provided staff to assist with collecting waypoints and provided equipment. An Indian Health Service intern collected waypoints and physical addresses. The USGS assisted with geocoding data and map design.

Outcomes

A primary outcome was the identification of the spatial distribution, which may contribute to increased asthma prevalence on the reservation. Second, the removal of two partially burned buildings in Poplar prompted the creation of maps using GIS (see this book's community mapping section in chapter 7). Finally, the project increased the application of GIS and how GIS may be used to understand complex public health problems.

Everyone benefits from projects that improve the health of a community. The maps created as a result of this effort validated previous efforts and tribal health concerns relating to asthma.

Recommendations

Consistency in measuring asthma prevalence in the AI/AN populations using GIS will prove important in the coming years. Chronic diseases are increasing dramatically, and improved GIS surveillance will help providers and patients better understand their causes. The unique aspect of using GIS to communicate public health information should not be overlooked. The saying "A picture is worth a thousand words" is much like a map created using ArcGIS Desktop: A map is worth indefinite benefits when it communicates the message tribal community members and leaders need to see.

More education is needed for asthma patients and their families to control environmental exposures. Additional study is warranted to determine the actual environmental factors associated with asthma in the context of their spatial relationships.

References

Agency for Toxic Substance Disease Registry (ATSDR). 2002. *Case Studies in Environmental Medicine, Asthma* (ATSDR-HE-CS-2001–2002), http://www.atsdr.cdc.gov/HEC/CSEM/asthma/index.html#authors.

Centers for Disease Control. 2004. "Asthma Prevalence and Control Characteristics by Race/Ethnicity-United States 2002." *MMWR*, 145–48.

Claudio, L., L. Tulton, J. Doucette, and P. J. Landrigan. 1999. "Socioeconomic factors and asthma hospitalization rates in New York City." *Journal of Asthma* 36, no.4: 343–50.

Mannino, D. M., D. M. Homa, C. A. Pertowski, et al. 1998. "Surveillance for asthma—United States." *MMWR CDC Surveillance Summary*, 1–27.

US Census Bureau. 2000a. "Census 2000." *American Indian and Alaska Native Summary File.* Census and Economic Information Center: Montana Department of Commerce.

US Census Bureau. 2000b. *Profile of General Demographic Characteristics for 2000: State of Montana and Fort Peck Reservation and Off Reservation Trust Land.* Montana. Table SF-1.

Assessing and mapping lead paint hazards on the Standing Rock Sioux Reservation

Dennis Painte, Environmental Program, Standing Rock Sioux Tribe
Joseph Kerski, Education Manager, Esri

The Standing Rock Sioux Tribe

The Standing Rock Sioux Reservation is a land of high plains grassland and buttes dissected by river valleys on both sides of the border between North and South Dakota.

Figure 7. The landscape of the Standing Rock Sioux includes grasslands, buttes, mesas, and rivers. Photo by Joseph Kerski.

Beginning with the Fort Laramie Treaty from 1869, negotiations led the way to the birth of the reservation on March 2, 1889. Greatly reduced from the original Great Sioux Reservation, Standing Rock contains about 850,000 acres, bordered on the east by the Missouri River, including the villages of Selfridge, Cannon Ball, Little Eagle, McLaughlin, and the tribal headquarters at Fort Yates. Its People are members of the Dakota and Lakota Nations, terms meaning *friends* or *allies*. The terms date back to the 1600s, when they were living to the east in the Great Lakes region, where the nearby Ojibwe People called the Lakota and Dakota *Nadouwesou*, meaning *adders*. French traders shortened and twisted the word to *Sioux*. The Dakota People of Standing Rock include the Upper Yanktonai or *Ihanktonwana* (*Little End Village*) and the Lower Yanktonai or *Hunkpatina* (*Campers at the Horn* or *End of the Camping Circle*). Yanktonai People today live primarily in the communities that are on the North Dakota portion of the reservation (Standing Rock Sioux Tribe 2010). The Lakota, the largest group of Sioux, are represented at Standing Rock as the *Hunkpapa* (*Campers at the Horn*) and *Sihasapa* (*Blackfeet*), and predominately live in communities located on the South Dakota portion of the reservation. Cattle ranching and farming are the predominant industries here; the Tribe also operates two casinos on its lands.

The Standing Rock Sioux Tribe (SRST) has a long history of using GIS for managing its lands and services. The Tribe maintains jurisdiction on all reservation lands, including rights-of-way, waterways, and streams running through the reservation. The tribal council passes legislation, makes budgets, approves of financial transactions, and makes major decisions affecting the Tribe, including managing property, business ventures, passing ordinances, entering into contracts, and making loans. It operates as a business but is guided by care and concern for the lands and the people inhabiting them. Indeed, its mission statement states that it is a "governing body empowered by the SRST Constitution committed to promoting an environment for the self sufficiency of all tribal members."

Lead paint health risks

People have been mining and using lead for thousands of years, even though it has long been known to be harmful to the human body. Lead (element Pb) adversely affects many systems in the body, resulting from a concentration of lead in the blood. This concentration can occur from "acute exposure," or a focused exposure in as short a period as several days, or from "chronic exposure" that occurs over a period of years. Adverse health effects include stomach problems, fatigue, moodiness, headache, anemia, joint or muscle aches, and difficulty in concentration, and if not treated, can lead to severe damage to the circulatory, nervous, skeletal, urinary, and reproductive systems, and ultimately death. Therefore, it is a health risk that must be taken seriously.

Lead has been used in fuel, pesticides, pipe soldering, and a variety of other commonly used products—including cosmetics. Lead contamination began to receive a great deal of attention in recent decades, particularly after it was shown that the most common route of contamination to humans was through an everyday substance—paint. Because of lead in paint, and also from its prevalence in soil and water, the US Department of Labor found lead poisoning to be the number one environmentally induced illness in children. At greatest risk are children under the age of six, because their neurological and physical development is occurring at a rapid pace and because they are often face-to-face with lead-based paint. At young ages, children are apt to place objects in their mouths or even eat paint chipping off walls. Lead poisoning can result in children having difficulties in studying and more permanent learning and behavior disorders. Because lead overexposure was also discovered to be one of the leading causes of workplace illness, the US Department of Labor became involved. In 1978, the federal government restricted lead in paint for residences, furniture, and toys to a maximum of 0.06 percent. In 2008, the EPA reduced allowable lead levels by a factor of ten–to 0.15 micrograms per cubic meter of air, giving states five years to comply with the standards.

Therefore, some lead is still permitted, and decades of widespread use mean that lead is inside virtually every person. The most common way that lead exposure is diagnosed is through measuring the blood lead level. This is typically measured in terms of micrograms of lead per deciliter of blood (μg/dL). The US Centers for Disease Control and Prevention and the World Health Organization consider levels of 10 μg/dL or above as "lead poisoning," but no known exposure level is considered "safe."

Addressing the problem

While much attention had been given to lead exposure, particularly in children, most of the focus has been on urban areas, particularly impoverished ones where older homes still contain lead paint. In Detroit, for example, more than half of the students tested in Detroit public schools have a history of lead poisoning, not just from their homes, but from the school buildings themselves (Lam and Tanner-White 2010). However, as Dennis Painte has discovered in Standing Rock, lead poisoning is also a problem in rural areas. Even though the total population of Standing Rock is only 9,000, with a rural population density below 10 people per square mile, lead is just as much of a concern here as it is in urban areas. The low population density presents its own challenges in terms of studying and addressing the problem and funding the necessary studies and treatment. Geography was fundamental to understanding, studying, and addressing the problem, and therefore, Dennis Painte had a vision and goal of using GIS as the tool to manage the problem from its start.

What gives Dennis and others hope in addressing lead poisoning is that unlike other diseases, in most cases, lead poisoning is preventable. It is preventable by decreasing the exposure to lead, through education of the public to the hazards of lead, and by direct action to eliminate or quarantine pipes, water, soil, paint, landfills, and other items that contain lead. Screening is an important part of preventive education and prescribing medicine. Treatment for those who show signs of lead poisoning include supplements of iron, calcium, and zinc, as well as more serious intervention such as surgery or chelation therapy, which involves injecting chemicals that form complex molecules with certain metal ions, which inactivate those ions. Another important part of its eradication is the removal of the source of lead.

Dennis received his GIS training through the Bureau of Indian Affairs (BIA) GIS program office in Albuquerque; his bachelor of science degree in environmental science; his work with the Standing Rock Environmental Protection Agency; and his coursework at Sitting Bull College, Oglala Lakota College, North Dakota State University, the University of North Dakota, and the University of Nevada Las Vegas.

Dennis says, "If a problem is identified and understood, then you can provide remedies." Because lead contamination takes place under certain circumstances, in certain places, and in certain demographics,

GIS was a suitable tool to help him understand the problem. Dennis began the project by making verbal assessments and visual assessments of every home in Standing Rock. He located the homes using a Trimble handheld GPS unit. Challenges in building the database include the time and miles of driving and walking needed to survey the individual households.

Next, Dennis contracted with a number of companies that perform risk assessments using an x-ray fluorescence (XRF) analyzer. This is a handheld device that analyzes metal alloys in the field, using miniaturized x-ray tubes. These devices isolate and measure trace amounts of lead from dust wipes that have been in contact with surfaces, as well as from soil samples and air filters. The XRF gun measures through to the substrate of the wall, such as brick or sheet rock. It can measure through twenty coats of paint that may cover up paint contaminated with lead far underneath the outermost layer. At the time of this writing, 287 houses were assessed by the Alaska-based company, with 14 more houses assessed by the Standing Rock Environmental Protection Agency, and 18 more are scheduled.

Along with the hazard assessment, Dennis coordinated a baseline assessment for children through the age of six using blood samples. He coordinated this through the Indian Health Service, whose staff performed the procedures after a parental consent form was explained and signed. Understanding that any project as personal as human health is sensitive and requires great delicacy, Dennis also coordinated lead education and outreach, including presentations about the dangers and hazards of lead and empowering people to take some steps to limit exposure. Recommended steps by individuals to reduce the blood lead levels of children included increasing the frequency of hand washing, increasing the intake of calcium and iron, discouraging children from putting their hands to their mouths, vacuuming frequently, and eliminating the presence of lead-containing objects such as blinds and jewelry in the house. Houses with lead pipes or plumbing solder should have those replaced with safe materials. Less permanent but cheaper methods include running water in the morning to flush out the most contaminated water and adjusting the water's chemistry to prevent corrosion of pipes. Lead testing kits are commercially available for detecting the presence of lead in the household.

There are about 2,900 homes in Standing Rock, and of these, 70 percent are Housing and Urban Development (HUD) homes. As the study continues, Dennis expects the GIS to reveal a random pattern of contamination, but it will be an important step in mapping and managing the program. Dennis works with a colleague doing similar work on the Fort Totten Reservation in North Dakota. He plans to receive training in operating the XRF devices, and will offer his services to other tribal governments to do this work on their lands. Thus, the experience gained at Standing Rock will help others.

One of Dennis' goals is to bring funding for mitigation so that the recommendations will be implemented. All of them cost money. For example, a single gallon of encapsulating paint costs ninety dollars. Therefore, he has secured two grants and is seeking additional grants such as a Healthy Homes Grant from the US Department of Housing and Urban Development (http://www.hud.gov/offices/lead/hhi/index.cfm), which addresses housing related hazards. He is hopeful that the reports and maps that he generates from the databases he is compiling using GIS will enable him to build a strong case as to why the grants are needed. Dennis is also encouraging the Tribe to build a recognized Lead-Safe Certification Program there. This program requires that contractors performing renovation, repair, and repainting projects that disturb paint in structures built before 1978 must be certified and follow specific work practices to prevent lead contamination. Thus, GIS has not only become a management and research tool, but also an advocacy and communications tool.

Conclusion

The vision of the Standing Rock Tribal Government is that it "strives to be a more effective, efficient, and visible government providing opportunities for our economy to grow through business development by educating our members, to enhance the health and wellness of the people of Standing Rock" (Standing Rock Sioux Tribe 2010). The lead paint study and mitigation using GIS is one more way in which the health and wellness of its People can be enhanced for a brighter future.

References

Lam, Tina, and Kristi Tanner-White. 2010. "High Lead Levels Hurt Learning for DPS Kids," *Detroit Free Press*. May 16. http://www.freep.com/article/20100516/NEWS01/5160413/High-lead-levels-hurt-learning-for-DPS-kids.

Standing Rock Sioux Tribe. 2010. "Tribal History." http://www.standingrock.org.

Connecting communities through primary, secondary, and informal education

GIS offers a powerful decision-making toolkit that can be used in educational administration, educational policy, and in instruction. GIS offers administrators at schools, colleges, and universities a way to monitor campus safety; map campus buildings, cable, and other infrastructure; route school buses; plan where and when to close schools and open new ones; and strategize recruitment efforts. GIS provides educational policymakers with tools to see patterns in educational achievement and where to implement new programs. In instruction, GIS in the hands of students helps them to understand content in a variety of disciplines, not only geography, but in history, mathematics, language arts, environmental studies, chemistry, biology, civics, and many more. So often, students feel that what they are studying in school has no relevance to their everyday lives. By contrast, GIS is used as a relevant, inquiry-driven, standards-based set of questions and tasks that let students solve real problems in real contexts. GIS fits in well with meaningful fieldwork and outdoor education (Louv 2005). GIS provides career pathways that are increasingly in demand in government, industry, academia, and nonprofit organizations. It helps students think critically about issues they care about and connects them to their own community. It does so in informal, primary, secondary, and university settings and appeals to today's visual learners.

The use of GIS also meshes well with Native ways of learning and knowing (Kerski 2006). GIS is a multimedia tool, incorporating sounds, photographs, videos, and links to web resources. It can help preserve Native languages through recorded voices in Native tongues and through the placement of Native names on the maps students create. Geotechnologies, along with biotechnologies and nanotechnologies, are the three most critically needed skills and fastest growing job markets identified by the US Department of Labor for the twenty-first century (Gewin 2004). As described in this chapter, students use GIS to generate and analyze data that is needed by tribal governments, and in so doing, contribute to the knowledge and well-being of their communities and people.

What is the relationship between land cover, precipitation, direction of slope, and wildfire occurrence on tribal lands? Why does this relationship exist? How does acid mine drainage in a mountain range affect water quality downstream? How will climate change affect global food production? With GIS, students explore the relationships between people, climate, land use, vegetation, river systems, aquifers, landforms, soils, natural hazards, and much more. With the flood of information available to students today, they need to be able to deal with uncertainty about data, to understand its limitations concerning errors and omissions, and to effectively manage it. GIS is a tool that provides holistic computer and management skills for students, a part of a new geospatial technology competency model recognized by the US Department of Labor (2010) that includes computer, personal, and organizational competencies.

Using GIS provides a way of exploring not only a body of content knowledge, but a way of thinking about the world (Bednarz 2004; Kerski 2008). These skills were identified as essential to K-12 education by the National Academy of Sciences (2006). The geographic perspective informs other disciplines. When epidemiologists study the spread of diseases, scientists study climate change, or businesspersons determine where to locate a new retail establishment, they use spatial analysis. In each case, GIS provides critical tools for studying these issues and for solving very real problems on a daily basis.

GIS in instruction also incorporates and depends upon fieldwork, which is critically needed for understanding and appreciating our world (Louv 2006). Students can gather locations with GPS and attribute information about tree species, historical buildings, water quality, and other variables on a field trip or even on their own primary or secondary school, college campus, and in after-school programs.

GIS-based questions begin with the "whys of where"—why are cities, ecoregions, and earthquakes located where they are, and how are they affected by their proximity to nearby things and by invisible global interconnections and networks? After asking geographic questions, students acquire geographic resources and collect data. They analyze geographic data and discover relationships across time and space. Geographic investigations are often value-laden and involve critical thinking skills. For example, students investigate the relationship between altitude, latitude, climate, and cotton production. After discovering that much cotton is grown in dry regions that must be irrigated, they can ask "*Should* cotton be grown in these areas? Is this the best use of water and other natural resources?" GIS helps students to act on their investigations, put their recommendations in place, and improve the quality of lives of people and the health of the planet. Students present the results of their investigations using GIS and multimedia. Their investigations usually spark additional questions, and the resulting cycle is the essence of geographic inquiry.

The stories in this chapter show how inquisitive students and innovative educators use GIS to enhance primary, secondary, and informal education. David Beres describes how students combine GIS, GPS, and fieldwork to understand and map the ways in which the Alamo Navajo People obtain water. Mark Ericson's students at the Santa Fe Indian School got out into the field and analyzed soil erosion and mapped agricultural fields in New Mexico. Lisa Lone Fight built a meaningful program rich in history, geography, biology, hydrology, and geotechnologies at the Wind River Native Science Center in Wyoming. Each story shows that GIS not only expands opportunities for students, but that it also provides benefits that tribal communities will feel the effects of now and far into the future through the contributions of these young people. In each case, GIS was more than just a technological set of tools. GIS built meaningful connections and partnerships between instructors and students; between tribal Elders and young people; between students and their communities; and between the past, present, and future.

Our world is constantly changing. These changes include those brought about by physical forces, such as erupting volcanoes, meandering rivers, and shifting plates, and by human forces, such as urbanization. On a local level, invasive weeds, wildfires, energy extraction, demographic shifts, and other forces change communities and lands. Students use GIS to understand that the earth is changing—to think scientifically and analytically about why it is changing, and then dig deeper: Should the earth and my community be changing in these ways? Is there anything that I should be doing or could be doing about it? This captures the heart of spatial thinking, inquiry, and problem-based learning. It empowers students, as they become decision makers, to make a difference in this changing world of ours.

References

Bednarz, S. 2004. "GIS: A Tool to Support Geography and Environmental Education?" *GeoJournal* 60(2): 191–99.

Gewin, V. 2004. "Mapping Opportunities." *Nature* 427(January 22): 376–77.

Kerski, J. 2006. "Earth Visions: GIS in Indian Country." *Winds of Change* 21(3): Summer.

Kerski, J. 2008. "The Role of GIS in Digital Earth Education". *International Journal of Digital Earth*. 1(1): 326–46.

Louv, R. 2005. *Last Child In the Woods: Saving Our Children From Nature-Deficit Disorder*. Chapel Hill, North Carolina: Algonquin Books.

National Academy of Sciences. 2006. *Learning to Think Spatially: GIS as a Decision-Support System in the K-12 Curriculum*. Washington, DC: National Academies Press.

US Department of Labor. 2010. Geospatial technology competency model http://www.careeronestop.org/competencymodel/pyramid.aspx?GEO=Y.

Students assess water quality and availability using GIS at Alamo Navajo Community School

David Beres, Science Teacher, Alamo Navajo Community School
Joseph Kerski, Education Manager, Esri

The Alamo Navajo community

Navajos call themselves *Diné* (the People), and make up a large subset of the Athapaskan (or Athabascan) community of Tribes inhabiting North America from Canada to Mexico. Other Athapaskan Tribes include the Jicarilla Apache, Mescalero Apache, Chiricahua Apache, Lipan Apache, Aravaipa Apache, Kiowa-Apache, Western Apache, Chasta Costa, Tutuni, Galice, Hupa, Kato, Eyak, Athapaskan, and Dene. Navajos are more closely genetically related to the Apache and culturally more closely related to the Pueblo Tribes, specifically the Hopi. Navajos were formerly known as the *Apaches Du Nabahu*, which was later spelled *Apache de Navajo*. Because of time, distance, climate, and landscape, a wide range of cultures now exist for the Athapaskan People.

The Navajo Nation covers a vast area of Arizona, Utah, and New Mexico, and most people are familiar with the main land mass, which contains nearly 14 million acres. Less well known are other land areas, all in New Mexico—the Ramah Navajo Chapter, with nearly 147,000 acres, the Canoncito (Tohajiileeh) Navajo Chapter, with nearly 77,000 acres, and the Alamo Navajo Chapter, which contains over 63,000 acres, or nearly 100 square miles, or 256 square kilometers. The Alamo Navajo lands lie in central New Mexico in northwestern Socorro County, adjacent to the southeastern section of the Acoma Indian Reservation, and nearly 2,000 people live there.

Because of its isolation from the rest of the Navajo Reservation, the Alamo Navajo People have maintained a unique traditional and linguistic heritage. The Alamo Band is the only Native American group of people who are a blend of Navajo and Apache Tribes. The Alamo dialect is Navajo and is commonly spoken in homes and in tribal government, religious, social, and cultural interactions, but is distinct in tone and phrasing from the Navajo spoken on the remainder of Navajo lands. Alamo Navajo is 220 miles southeast from the Navajo capital of Window Rock, Arizona, and is also isolated from other communities. Its main tribal community lies thirty miles northwest of the largest town in the area, Magdalena, which has only 1,100 residents, and fifty-seven miles from Socorro, which has less than 10,000 residents.

The main Alamo Navajo community lies along Highway 169 just north of the Cibola National Forest, but residents also live in other sections of the reservation. The Navajo word for land or geography is *Ke'yah Baa Hane'*. The physical geography of the Alamo Navajo is beautiful—semi-arid rangeland dotted with piñon and junipers, rolling hills, badlands, volcanic rocks and lava flows, and mountains. However, isolation from the main Navajo reservation and from surrounding communities presents its challenges, including socioeconomic and educational gaps between Alamo and the main Navajo reservation.

Alamo Navajo Community School

Given its isolation, it is perhaps understandable that the Alamo Navajo had no school of its own until the 1930s. This was a school operated by the Bureau of Indian Affairs (BIA) but soon closed in 1941. For the next forty years, children from Alamo Navajo had to be sent to boarding schools as far away as Santa Fe. The situation was eased a bit in 1957 when a BIA dormitory was built a bit closer in Magdalena, thirty miles southeast. Still, students living there had to attend the public schools of Magdalena, rather than in their community. Finally, many dreams came true when the Alamo Navajo School Board was created in 1979. Drawing from the Indian Self-Determination Act of 1975, the school board opened a community school on October 1, 1979. Originally a K-8 school in four portable buildings, it soon expanded to a permanent building of 54,000 square feet and twelve grades, including classrooms, laboratories, a library and media center, a gymnasium, athletic fields, and cafeteria. The school board went on to operate an Adult Education Program and an Early Childhood Center. They even went beyond traditional education to create an Indian health center, a new multimillion-dollar wellness center, a roads department, and a technology department. Currently, 350 students attend Alamo Navajo Community School. The school is accredited by the New Mexico Department of Education and the North Central Association of Colleges and Schools.

The GIS Program at Alamo Navajo Community School

The Alamo Navajo Community School's GIS Program is housed in science, but it has disciplinary ties to technical and vocational education, life skills, mathematics, and social studies. The critical thinking and analytical skills offered by GIS mesh well with the mission of the school to create and maintain an open learning environment that encourages and helps students to work toward their highest potential in effective communication, creative and critical thinking, self direction, responsible learning, discovering their talents, demonstrating compassion and dignity, and contributing to the global society. The key workforce skills offered by GIS fit well with the school's vision to prepare its students to have unlimited choices after graduating from high school. The high school program at Alamo Navajo Community School makes a point of focusing on language—reading, phonics, word recognition, spelling, vocabulary, grammar, comprehension, and fluency, in both English and Navajo. Other subjects besides language arts include radio journalism, social studies, science, mathematics, art, business education and life skills, and technical and vocational education. It also offers an after-school program in arts, crafts, and other educational activities.

On a broader scale, the GIS Program also has close ties to the school board's mission and vision. The school board's mission is to offer the necessary services and resources to empower community members to attain good health and self-sufficiency through educational excellence and to carry out the mission in the spirit of Indian self-determination and local decision making in the planning and administration of its programs. Through the use of GIS, students can contribute to the future of the Alamo Navajo and the decisions they need to make about housing, health, water, energy, education, and other issues. Furthermore, using GIS and GPS in the classroom and in the field empowers students because they are using the same tools and data as practicing planners, hydrologists, archaeologists, and others.

Through its layering of rich data and its holistic approach to conservation and other natural resource issues, GIS meshes well with Native cultural and educational values. The goal of using spatial tools is to create more sustainable living spaces. Empowering students and linking well with Native ways of learning and knowing supports the vision of the school board—for Alamo community members to attain physical, spiritual, and emotional well-being; community and individual self-sufficiency; and realize harmony between tradition, cultural values, and the mainstream environment.

David Beres is the science instructor at the school. He completed his fifth year at Alamo Navajo Community School during 2010. He followed quite a long migratory path to Alamo Navajo, as he formerly taught science and mathematics in Alexandria, Egypt. He has a master's degree in social work and has worked with the elderly population but prefers working with students. Now in his tenth year of teaching, he has seen many technologies come and go and does not adopt technology just for technology's sake. He values the use of GIS in education for several reasons. First, GIS fits in well with the hands-on learning style that the students have. They don't want to be lectured to all day—they want to get busy and *do* things. In his chemistry classes, students experiment with many elements and compounds. In his biology classes, students examine life forms with microscopes. In astronomy, Mr. Beres requires students to find specific constellations. In Mr. Beres's words, "everything in mapping is a hands-on project" and therefore, GIS is a perfect match for the way he teaches. Each topic he teaches is interdisciplinary. In order to know something about topic A, students may need to know something about topics B and C. They then investigate those topics. GIS is also an interdisciplinary learning tool. In working with GIS, students may need to know something about landscape, settlement patterns, the electromagnetic spectrum, and if so, they study these topics. During spring 2010, Mr. Beres used ArcGIS Desktop. He also used ArcGIS Online, especially important in a region of the world where few digital basemaps exist. ArcGIS Online enables students to quickly map the data that they collect with GPS units on the school campus or in some other location. It also allows students to investigate demographics, rivers, and other phenomena in the region. Mr. Beres learned how to use GIS as an educator. He has taken a course through the National Science Foundation and has learned about GIS via some Esri Press books, but he has largely learned GIS while teaching it to his students and learning from his students.

Mr. Beres believes that GIS also fits in well with the new "expeditionary model" that the school is following. The expeditionary model is in part the school's response to increasing pressure from the local Bureau of Indian Education school district to raise performance standards and skills, something every school is facing these days. The expeditionary model is an advanced method of study that shares some characteristics with the International Baccalaureate (IB) Program. All teachers in the program build their course content around a single theme. Projects are open ended.

GIS also complements the "what the students see" approach. For example, Alamo Navajo students spend much less time studying glaciers in earth science, simply because glaciers are not a part of their world. They spend much more time on desert and volcanic processes.

Mr. Beres chooses technologies and methods that will foster peer mentoring—students teach and learn from each other. Students using GIS in Mr. Beres's classes exhibit a great deal of peer mentoring as they help each other with gathering spatial data, uploading their GPS waypoints and tracks, analyzing the resulting spatial patterns, and producing GIS-based maps.

The principal and superintendent of the school district trusted Mr. Beres when he enlisted their support for the use of GIS at the school. One thing that helped was that the computer hardware was already in the laboratory. Two years ago, when the Alamo Community Wellness Center in Magdalena, New Mexico, opened, the staff learned that Mr. Beres worked with GPS and GIS. They wanted him to construct a geocaching course that fit in with their aims of providing fun and interesting fitness activities. At the time, Mr. Beres's class owned two GPS units, but he was able to convince the wellness center to purchase twelve more of them. The students have mapped a wide variety of features and phenomena with them.

Water quality project

In part because of its semiarid environment, and in part because of the poor quality of its infrastructure, the Alamo Navajo have long had a major problem with water on their land. The water system frequently breaks down, depriving isolated residents of water and creating a real concern. The Alamo Navajo rely completely on groundwater on the tribal lands. In some areas, windmills pump groundwater to the surface. High winds often destroy the windmills. A push is underway to modernize the entire water delivery system. The school board was asked to manage the water system. According to Mr. Beres, the Alamo Navajo Tribal Government is not using GIS but relies on paper maps. In addition, the tribal government is too small in staff size (six) to take on the project. Therefore, the school board asked David Beres to make a map of the reservation to assess water quality and availability.

Mr. Beres immediately turned to his students. He had four of them in his GIS class at the time, and they divided the reservation into quadrants. He set up the course as a summer work project. First, they identified all the homes within the quadrants. They held a meeting at 8 a.m. each morning to discuss concerns and review overall goals. They then drove to each home in the school vehicle, took a waypoint with their GPS receiver, and interviewed the occupant about the water quality and availability. Collectively, they

Figure 1. High school students collecting data with GPS receivers outside the Alamo Navajo Community School. Courtesy of Alamo Navajo School.

Figure 2. Wagon Mount Well is one of the typical water wells at Alamo Navajo, which was visited, geocoded, and tested by students using probes and GPS. Water is pumped from the ground into the tanks in the background and then to the homes of the people living in the area. Buffers are run to determine how far each household is from water sources. Courtesy of Alamo Navajo School

visited up to 96 percent of the 500 homes on the Alamo Navajo tribal lands. The students created a database of water users, discovering how people are getting water from the water system. It improved the maintenance of the system and the accuracy of billing. More importantly, it helped identify homes where water pipes or the water meter was leaking or where the water was unsafe. Nine major wells existed on tribal land. There were pumps on several of them. The Navajo Environmental Protection Agency found the data to be extremely helpful. The project showed administrators that the Tribe was tackling an issue of concern and could use the results to ask for additional funding to improve the water system.

The project was successful but progress slowed when the student summer class finished. Mr. Beres and his current students still work on the project, assessing and mapping data.

Benefits of using GIS at the Alamo Navajo Community School

GIS has been so valued by students that some have asked to access the computer lab even after they have completed Mr. Beres's GIS class. They will stay during lunchtime and after-school hours to work on projects and to explore datasets. Students are engaged in the learning process, take ownership of their own education, and use data and tools to investigate real places and issues. They also, most importantly, use their heads: GIS is a thinker's tool, fostering critical analysis and hypothesis testing. Through the water quality study, students used GIS in a project that served all tribal members as well as the tribal government. Research has shown that meaningful service learning projects connect students with community issues, promote good citizenship, and may encourage students to return to the Alamo Navajo to serve the Tribe in other areas. Internship opportunities for students in GIS are extremely limited, but Mr. Beres has paved the way for one student to do GIS work with the local US Forest Service office. Using GIS also encourages students to pursue higher education: The school's GIS class, like other technology-based classes at the school, is part of an articulation agreement with Southwestern Indian Polytechnic Institute (SIPI) in Albuquerque. The school teaches the first few technology classes, and these courses transfer to SIPI where the students pick up where they left off at Alamo Navajo Community School. In the fall of 2010, the GIS Program will become a part of the expeditionary program at the school.

The benefits of GIS are not limited to the students. The water quality study has benefited the entire Alamo Navajo Tribe. Students took a critical issue of concern—water, which impacts the lives of each person—and improved the situation. They benefited the lives of the Alamo Navajo People now and those of future generations.

Combining fieldwork, environmental issues, GIS, and GPS at the Santa Fe Indian School

Mark Ericson, Environmental Science and Technology Instructor, Santa Fe Indian School Community Based Education Program

GIS and GPS through community-based education

Since 1997, students in the Santa Fe Indian School Community Based Education (CBE) Program in Santa Fe, New Mexico, have used GPS and GIS technologies to supplement their fieldwork. Over the years, partners in the Pueblo communities have engaged the students in a broad range of field-based research and projects that address important issues the Tribes are trying to address, including aquatic habitat assessment, wetlands mapping, vehicle caused erosion, mapping of dump sites, and rangeland studies. Combined with interdisciplinary studies of current, relevant, and important community issues, these experiences have influenced many students who have been through the program to pursue post-secondary education in related science, technology, engineering, and mathematics (STEM) fields. Indeed, many have gone on to work in their own tribal communities, contributing their GPS, GIS, and other skills and talents to solving problems and helping their Tribes.

The CBE Program is designed to allow students and staff the time to travel to their partner communities to conduct fieldwork on a regular weekly basis. Our community partners are the environmental departments of some of the Pueblo Tribes near Santa Fe. The types of projects are decided by the partners in a planning process at the beginning of the school year, and often there are situations that arise spontaneously that we are able to take advantage of because of the flexibility of the schedule. GPS-based collection of study sites and GIS analyses are standard parts of the project documentation process, and are often products that the partner community needs. Students use some of their time at school to manage and process the data and produce these mapping products.

Students are exposed to, and learn about, the variety of environmental issues facing Tribes. While only some of the students are from the Tribes we work in, they learn about issues that their own Tribes must deal with, and that there are different ways of addressing these issues. As the program has developed, our community partners have realized the value of the students' contributions. Often, the work the students are asked to do is unprecedented and produces needed baseline data. The emphasis has been on involving the students in producing valuable information that the Tribe can use. This is information that the environment department needs but doesn't have the time, personnel, or funding to commit.

Erosion, agricultural fields, and the unexpected

Sometimes, there are unexpected and fortuitous outcomes from the students' work. One example was the mapping of the edges of a large erosion headcut, an erosion feature that was representative of significant topsoil loss in an area that had suffered from overgrazing by livestock.

Figure 3. Students in the field mapping one of the headcuts that is causing severe soil erosion on and off tribal lands in the region. Courtesy of Marc Ericson

Chapter 6: Connecting communities through primary, secondary, and informal education | 97

Gathering coordinates via a GPS of the edges of the headcut banks was a routine application of the technology during a field trip where the students learned from the rangeland specialist about erosion issues in general, how they occur, the significant threat to community sustainability that these erosion features represent, and methods that are used to cease their growth and begin the slow process of mending the land.

Back at school, the GPS files were downloaded and overlain on the available 1-meter resolution digital orthophotographic files for the area using ArcGIS. At this resolution, even though the satellite geometry that day was excellent and the Trimble GeoXH data is usually accurate to within a foot, and certainly to within a meter, the result was disappointing because the erosion feature itself was not very clear in the digital orthophotography, so no alignment, or matching of the GPS data with the headcut, could be discerned.

Needing higher resolution orthophotographic coverage, a search was made and incredibly, because of the importance of this area to the Middle Rio Grande Conservancy District, it was discovered that 6-inch digital orthophotographic data for the area existed. Once the student GPS files were overlaid on the new image, it became instantly apparent that the extent and rate of growth of the headcut could be

Figure 4. Results of student analysis of erosion rates on a typical headcut as mapped in the GIS environment. Courtesy of Marc Ericson

measured. Using ArcGIS measuring tools, including the knowledge of when the digital orthophotograph had been taken and when the GPS data had been collected, it became possible to measure the rate of headcut growth and develop an approach to estimating the rate of soil loss.

In itself, this was very compelling and functional information for calculating the rate of topsoil loss, but this information gave the tribal rangeland specialist useful knowledge for making the case for the urgency of implementing strategies for stopping the headcut growth. This approach has led to a methodology and desire to involve students in the baseline mapping of other headcuts that are eroding the landscape, and developing methods for determining erosion rates and quantifying soil loss as a way of prioritizing application headcut cessation efforts.

Another example of how student fieldwork has helped a Tribe involves the GPS and GIS mapping of agricultural fields. An approximately fifty-acre area of reservation that had been fallow, and which had received help from the Bureau of Reclamation to install new irrigation equipment, was graded and bermed. Students used GPS units to map the plot perimeters and areas, the turnouts, pipelines, valves, trees, and other features. Students in the CBE agriculture class then took soil samples that were sent to New Mexico State University for analysis, and the analysis results were linked to the points in the fields. The overall GIS with plot areas and soil analysis results and recommendations were used by the Tribe to help allot the farm plots to community members. This approach was developed as a methodology that could be applied to larger areas of the same reservation that are being readied for farming and for farming areas on other reservations.

In addition to fieldwork that is designed to expose students to real environmental issues and involve them in data gathering and management, students in the CBE Program are given the data and GIS tools to create comprehensive maps of their reservations and surrounding areas. Most students enter the program with little knowledge of their reservation legal boundaries and the spatial relationships between natural features, such as waterways and bioregions and human impact sites. We show them how to use digital elevation models (DEMs) processed with the ArcGIS Spatial Analyst extension to derive watershed basins and stream channels and then convert interest areas to triangulated irregular networks (TINs) for visualization with the ArcGIS 3D Analyst extension. Along with 1-meter digital orthophotography (and 6-inch if it is available), and other available shapefile and other data showing roads, habitation, farming fields, land ownership, and other features and attributes, the students develop comprehensive maps of their reservations and surrounding areas. They learn who their neighbors are and their legal jurisdictions so that they are better prepared to protect their natural resources in the future when they begin to fill positions of authority within their communities. Currently, work is being done to develop this process to include possible reservation habitats of listed and candidate threatened and endangered species.

Since water is of such crucial importance to Tribes, practically and culturally, the CBE Program helps the students develop a better appreciation and understanding of the nature and extent of the watersheds that feed their reservations. Students learn about watersheds in general, as an overarching concept, which can be used to integrate math, science, and social studies. Watersheds are examined to learn and inventory their intrinsic attributes (their biological, hydrological, and geological attributes) and the instrumental uses humans have made and currently make of them (the habitational, agricultural, livestock, forestry, industrial, transportation, and recreational ways in which the intrinsic attributes are

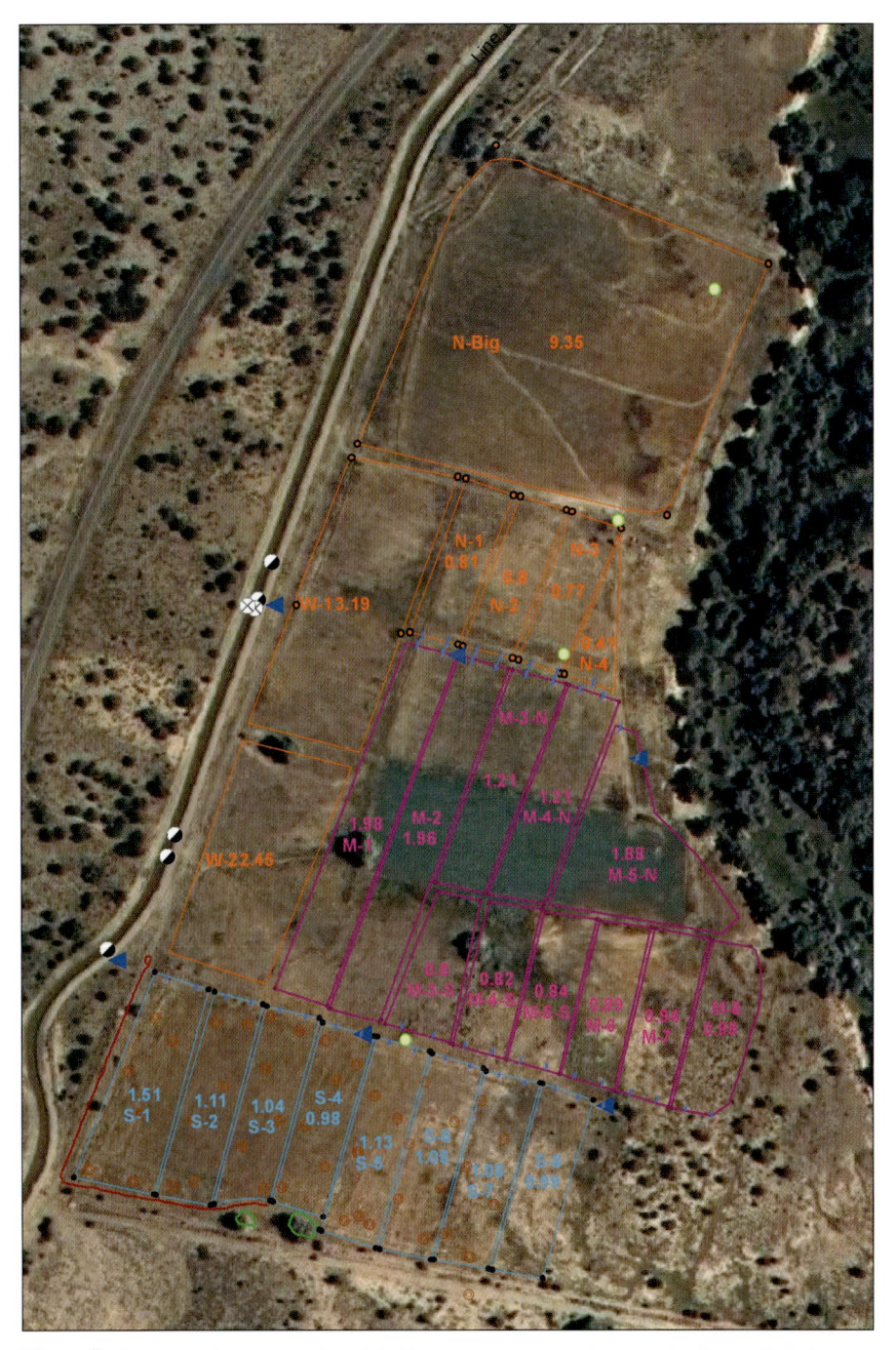

Chapter 6: Connecting communities through primary, secondary, and informal education | 99

Figure 5. Example of the agricultural field mapping that students at the Santa Fe Indian School completed. Courtesy of Marc Ericson

used). In this way, students have a basis for developing an understanding of, and assessing, the health of watersheds that are important to them, and the processes that might be used to develop watershed management plans with the other watershed stakeholders to ensure that the watersheds maintain a healthy balance into the future, particularly with concern for changes in precipitation and temperature patterns that are already manifesting as climate patterns change. GPS and GIS technologies have become core components of this process.

There are many ways and levels in which students can use GIS technology in the context of learning about community issues. Through the interdisciplinary study of important community issues, combined with meaningful field experiences and training in the use of technologies like GPS and GIS, young community members can make valuable contributions to their Tribes. This approach also reinforces and enriches the learning of math, science, social studies, and communications concepts. Students have real experiences on which to base their interests, and for many of the students who have been through the CBE Program, these experiences have been compelling enough to influence them to pursue related studies and work after high school, at the university level, in the private sector, and within their own communities.

In addition to the crucial support of tribal and school leadership, the Santa Fe Indian School's Community Based Education Program has received generous technological and funding support from Intel and the US Department of Energy.

Indigenizing, contextualizing, and applying GIS technology and spatial analysis to meet the needs of Native students and communities: A view from the sky

Lisa Lone Fight, Mandan, Hidatsa and Arikara Nation Indigenous Scientist and Educator
http://www.earthlodge.net

The world of the Mandan, Hidatsa, and Arikara People is alive with stories. They are our maps as we journey through life, and they are powerful. In a sense, our stories are the aerial view of our lives and reality. They exist intertwined as elements of a dynamic cultural landscape. Some are lived stories: The flooding of our reservation in the 1950s and the relocation of my family. My grandmother, Buffalo Bird Woman, the remarkable scientist and author, whose book is still in print one hundred years later. My mother, the culture bearer and journalist, composing a column while flying in a bomber with the Air Force. My father, biologist, teacher, and Tribal Chairman, playing the drum and singing Sitting Bull's battle song prior to addressing Congress and negotiating reparations from President Ronald Reagan. And a thousand others.

Our lived stories are our home. They are warm and nearby—a close-up view. When we climb higher and our view becomes more remote, we see that these stories are only points that exist within an

Chapter 6: Connecting communities through primary, secondary, and informal education | 101

extraordinary lattice. Because we are wealthy—because the chain of our oral tradition has no broken links—our stories extend back to the beginning of creation, a time describing existence before the universe was completely formed and before we were fully settled in "this" world. These are the stories we tell in winter. They are our means of making sense of and exploring the world, and they aid us in understanding not only how the world worked/works but "how it was meant to work," and perhaps most importantly, who we are and "who we are meant to be."

I have always existed immersed in our stories, and while all these have, in a sense, "mapped" my life, there is one story in particular that defines, empowers, and compels the scientist in me. This story is about a woman who climbed into the sky. In one part of the story, the woman becomes homesick. She makes a hole in the clouds and sits gazing down at the land and the people below. This has been a vision that has returned to me over and over throughout my personal and professional life. "What was she seeing?" I asked myself. The buttes, valleys, buffalo herds, tall grass prairies. "What would she see today?" A lake where there was once a free flowing river, houses instead of earth lodges, cattle instead of buffalo. And most importantly to me, "What would she think of this new world she looked down on?" What would her "analysis" be, and how could that help her/our/my People?

It is commonly known that Native People are tied to geography; we are a People of place. The very nature of being "Indigenous" demands a location to be Indigenous to. We are also, however, People of space, image, and time; this is lesser known. We constantly seek perspectives and knowledge of the world that explain it and the beings within it. Our heroes go to buttes and mountains and often, when the need arises, take to the air. Remote sensing and geospatial science, the seeking of knowledge from a distance and placing it within a landscape, pervades our culture. It has been our business for thousands of millennia; in truth, it has been our business since the time that woman first looked down from the sky.

Given our cultural context, it is not surprising that the history of GIS, indigenous People, mapping, and sensing has no clear beginnings. It does however, have critical junctures, and one of these world-changing events for the Mandan/Hidatsa People was the meeting of Lewis and Clark with my grandfather Sheheke Shote. Lewis and Clark were "colonial" mapmakers; that was their career and mission. Their colonial mapping was the mapping of "claiming" and "owning." While they made use of the indigenous maps—mental, literary, historical, and cultural that existed long before their "Corps of Discovery"—they did not comprehend them because indigenous maps exist as the collective knowledge and wealth of Native People. The miracle is that these maps still exist today, and their existence is vital, political, scientific, and enormously and exquisitely powerful. The beauty of being an indigenous educator and scientist in the twenty-first century is that when indigenous mapping and new GIS and remote sensing technologies are combined, extraordinary things begin to happen.

Historically, maps have been used to define away indigenous rights, and while indigenous Peoples can claim much of the planet as ancestral territory, they actually occupy and manage a much smaller percentage in a political and economic sense. It may seem obvious but it bears repeating that *all* land ownership is defined by maps, and changing the map changes the ownership. Even today you will see reservation boundaries in the United States disappear and reappear on maps, depending on the aims of politicians, the requirements of administrators, and the accuracy of mapmakers. While it may be argued that these are "bad" maps, the only answer is to have "better" maps. Increasingly, when borders, boundaries, or resources are contested, GIS scientists are courted for validation. This provides the opportunity for contemporary methods of remote sensing and GIS to be new tools of empowerment for indigenous People and their allies. The only problem is that the number of indigenous People who command these technologies can be counted on a few sets of hands.

In my family and Tribe, images have long been understood to have extraordinary power. This power is realized by all Mandan, Hidatsa, and Arikara People when they view the relatively famous picture of Chairman George Gillette as the legislation was signed that would flood our reservation in 1951 (see http://publications.newberry.org/lewisandclark/newnation/ranchers/flooding.html).

This photo is usually accompanied by aerial photographs of the waters of Lake Sakakawea (then still the Missouri river) consuming the ancestral homes of the Mandan, Hidatsa, and Arikara People (see http://clui.org/sites/default/files/imagecache/clui-image/clui/post_images/0194-14-aerial-view-of-elbo.jpg).

These images were an early demonstration of the power of remote sensing. The emotional impact is extraordinary, but the scientific value is beyond price. The power of GIS is to show what has, what is, and what could happen to our lands. The politics of the twentieth century brought flooding and land seizure. What will the politics of the twenty-first century bring as the courts increasingly "define away" our boundaries and change our maps based on the advice and testimony of scholars, historians, lawyers, and, increasingly, GIS scientists? How do we prepare ourselves, our communities, and our children for this future? Education has always been an answer, and GIS education is alive with possibilities.

My own career in GIS leapt forward in tandem with my career in education. I had the good fortune and blessing to serve as the director of the Wind River Native Science Field Center (WRNSFC). Based on needs and concepts voiced by indigenous Elders and communities, three Native Science Field Centers were implemented by Hopa Mountain and funded by the National Science Foundation to foster informal science education from an indigenous perspective. One was placed at Oglala Lakota College in Rapid City, South Dakota, a second at Blackfeet Community College in Browning, Montana, and the third was the WRNSFC, which was located on the Wind River Reservation in Wyoming, home of the Eastern Shoshone and Northern Arapaho People. It was quickly apparent that with the Wind River Reservation encompassing almost two million acres, GIS technologies and remote sensing were important tools for providing critical information to the People of the Eastern Shoshone Tribe and Northern Arapaho Nation. When approached with the concept of training students in their use, the support from both Tribes and communities was overwhelming. This occurred at a crucial time because one of the most significant issues on the reservation was the extent of land-based tribal jurisdiction.

Jurisdiction, like ownership, is always and inextricably tied to maps. When looking at maps of the Wind River Reservation, you will find the reservation generally highlighted as a square in West-Central Wyoming.

Generally this square is displayed as intact, but increasingly it is the tendency of some mapmakers and politicians to represent it with a large wedge removed from the right side. This has profound implications for all people involved but most significantly for Eastern Shoshone and Northern Arapaho children. They are the ones who will inherit one vision/map of their reservation and sovereignty or another.

At the WRNSFC, we looked to the community members to express their preferred ways to help their young people become knowledgeable about and gain control of the processes of "definition" represented by mapping. Based on the community response, GIS was folded into the action plan utilizing a learner-centered community-based paradigm. The research agenda was determined by the students

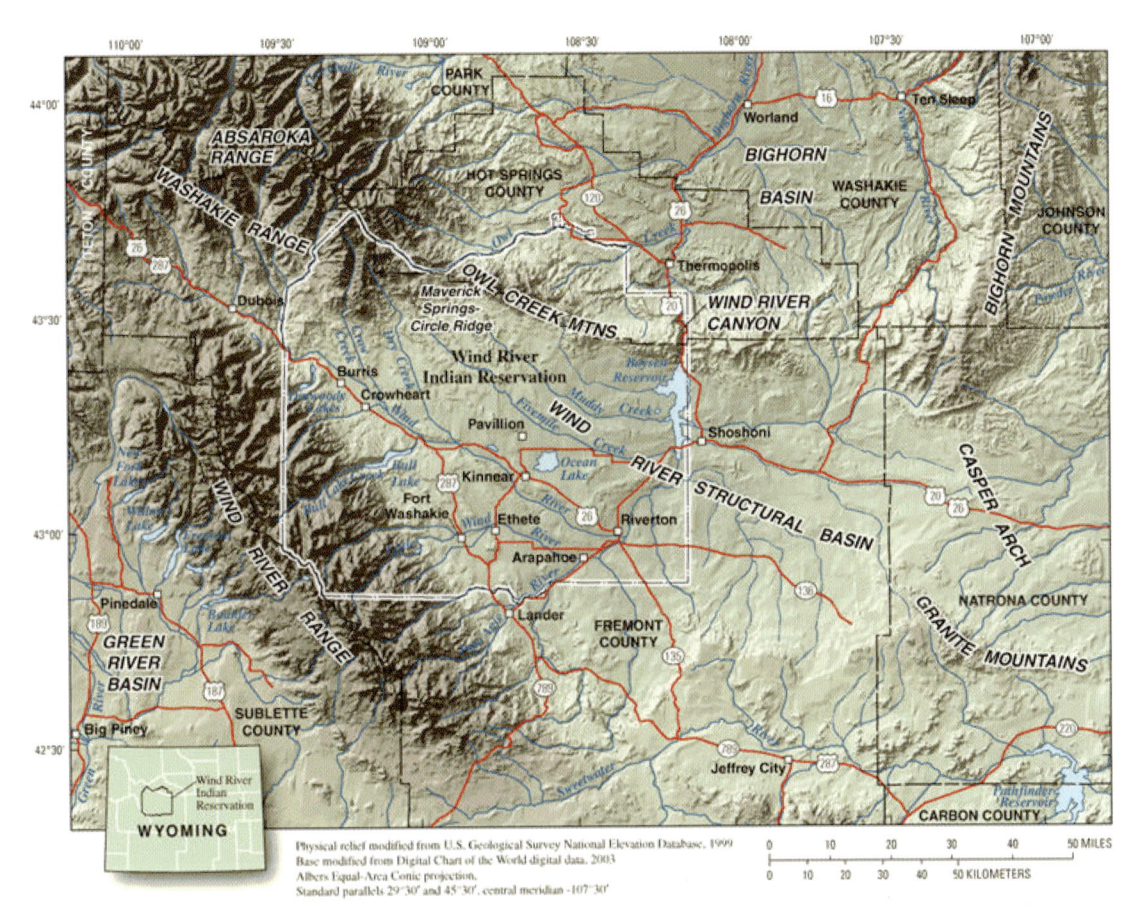

Chapter 6: Connecting communities through primary, secondary, and informal education | 103

Figure 6. Wind River Reservation in West-Central Wyoming. Courtesy of US Geological Survey, http://pubs.usgs.gov/sir/2005/5027/

in consultation with the Elders, teachers, and community members. Students were trained in mapping using a model that incorporates stories and spatial thinking, known as a "Stories and Spaces" model. Students used color touch screen Garmin Oregon 400 model GPS units and as a generation weaned on video games, it took them only moments to figure them out. They were guided in the use of the GPS units by educators and tribal GIS experts and required to do fieldwork where they used the units to map geologically, culturally, environmentally, and politically important locations and resources on the reservation. The students worked in conjunction with the Tribal Environmental Quality Department and their maps of their reservation and linked all data with rich cultural content provided by tribal Elders describing the locations of cultural resources such as traditional plant medicines and traditional hunting routes. These activities were framed explicitly as a method for "claiming" reservation spaces, and the project was designed in a self-sustaining, open-ended manner so it could be added to at any point there was student or community interest. The outcome was remarkable—young Native American students had been trained in the technology of mapmaking and were defining their own indigenous

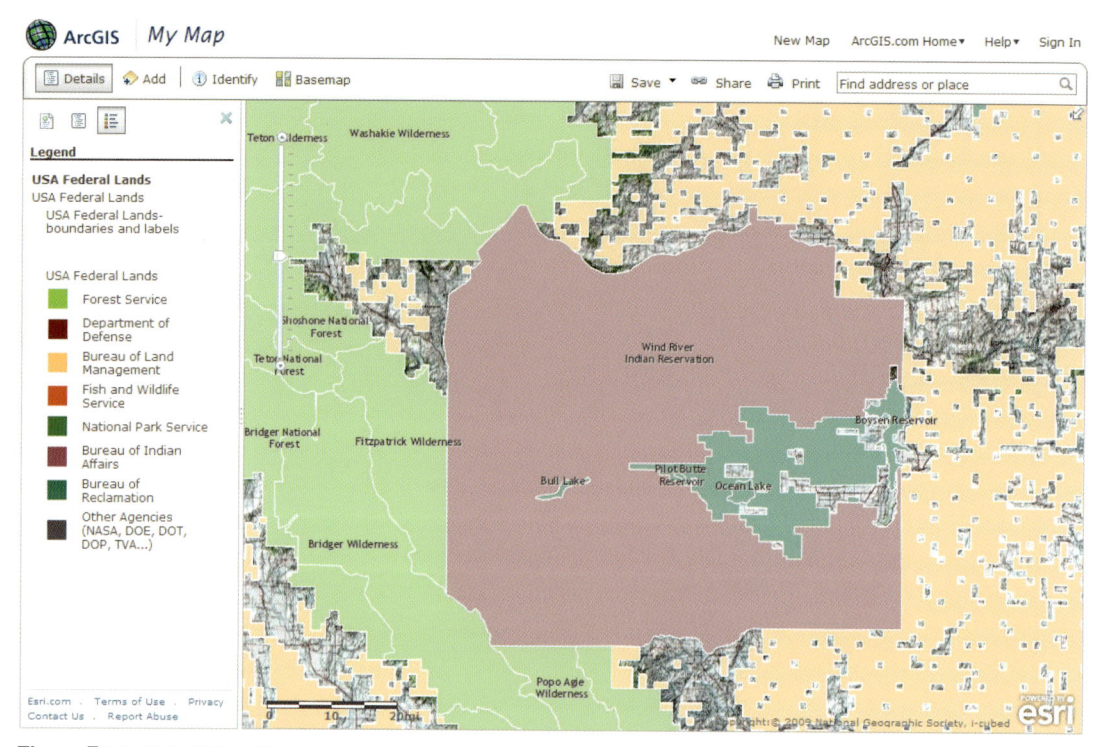

Figure 7. The Wind River Reservation shown with diminished tribal jurisdiction on its eastern side. Courtesy of National Atlas, Esri. Source: Federal lands from nationalatlas.gov shown via ArcGIS Online.

spaces and validating them with the stories of Elders. The students who would be the caretakers of the land used spatial science technologies and elder wisdom to explore the very lands they will be asked to steward.

The WRNSFC project was an example of indigenous mapping at its best. True indigenous mapping explores and gravitates toward the intersections of culture, history, tradition, geography, education, and community empowerment. It employs participatory mapping to empower indigenous People to make their own maps and interpret them. Personally, I use a process I have termed "Respecting Indigenous Participatory Spatial Sovereignty" (RIPSS). This GIS/remote sensing knowledge generation process is grounded in the following understandings:

- Indigenous People (explicitly including Elders, students, and tribal decision makers) must be involved in the collection, analysis, and application of indigenous spatial knowledge.
- The knowledge of indigenous Elders and community members must be respected and validated throughout the process.
- Indigenous spatial knowledge must be contextualized within oral traditional and cultural history.
- Indigenous spatial knowledge must not be used to harm indigenous communities environmentally, historically, economically, spiritually, linguistically, or culturally and is generally considered the "property" of the community.

This process provides a framework for placing GIS/remote sensing at the service of indigenous communities in ways that are useful, respectful, and empowering.

Working as a spatial science professional, I often think of the woman looking down from the clouds and feel the reach of her vision. Our People have long understood the power of remote sensing and GIS. Using a small portion of this historical and timeless perspective joined with the latest technology allows indigenous People to predict, document, and in the best of worlds avoid the devastation and loss that has occurred on my reservation and many others. This is a vision of a Mandan, Hidatsa, and Arikara woman, scientist, and educator indigenizing remote sensing and GIS. What the next generation of Native educators, students, scientists, and leaders will accomplish is unknown, but they must have the best tools and training to achieve it. It is their right and legacy to have a view from the sky that maintains, celebrates, and protects their indigenous earth.

ved by Joseph Kerski

Building career pathways through higher education

The stories in the last chapter illustrate how effective GIS can be with young students at the primary and secondary levels of education. The stories also showed that GIS can be used easily in formal school settings and informal ones, such as after-school programs, and in museums, science centers, libraries, and clubs. GIS can also be used in instruction, educational policy, and in higher education administration. In higher education, students using GIS can gain key workforce skills in desktop and web GIS software and programming languages, spatial analysis, spatial statistics, cartography, and field data collection techniques, including using GPS and scientific probes. But students also gain skills in areas extending beyond GIS technology, such as data management, communications, and research. These skills will serve them well, even if they do not pursue a career in GIS. Because GIS is becoming an important tool for physical and social sciences, engineering, health, and many other fields, working with geospatial technology is becoming a fundamentally important part of all higher education. For the instructor, the use of GIS offers opportunities to partner with faculty in other departments across campus or even on other college or university campuses.

The stories in this chapter show how students and instructors are using GIS to enhance higher education. Tammie Grant explains how students through summer workshops at Salish Kootenai College (SKC) used GIS, GPS, and remote sensing technologies to study grizzly bears, wildfires, and landscapes. The faculty at Haskell Indian Nations University discusses how students there have mapped and analyzed topics as diverse as plants, accessibility on campus, and ice sheets in Antarctica. Allyson Kelley and her colleagues talk about the partnership that enabled a series of community mapping workshops on four reservations in Montana. Just as in the world of primary, secondary, and informal education, these stories from higher education illustrate that GIS expands opportunities for students and faculty and provides long-lasting benefits for tribal colleges and communities.

Investigating the world through GIS and remote sensing at Salish Kootenai College and beyond

Tammie Grant, Educator; and Joseph Kerski, Education Manager, Esri

Description

Tammie Grant's longstanding passion for teaching, geology, and geospatial technologies led her on a journey that has enriched specific environments and also the lives of many individual Native students. Her journey began with a geology degree, and she continued working on planetary geology at NASA's Goddard Space Flight Center in Greenbelt, Maryland. She moved from planetary geology to vegetation mapping and worked on the Global Inventory Modeling and Mapping Studies (GIMMS) Project (http://gcmd.gsfc.nasa.gov, search for GIMMS), which provides scientific earth systems research through satellite remote sensing. One of the datasets produced from GIMMS is a global measure of normalized difference vegetation index (NDVI) that covers a twenty-two-year period, used to monitor photosynthesis and energy absorbed and released by the earth's land surface. With the project's partners, Tammie mapped very large areas to investigate areas of tropical deforestation, including the Congo, the Amazon basin, Borneo, and Indonesia.

Tammie's desire to work with Native Peoples in the field of geospatial education led her to approach NASA with a remote sensing and GIS education proposal focused on Native Americans. It was suggested to Tammie earlier through NASA's Minority Program that SKC (http://www.skc.edu/) would be a great place to start. The college had been using GIS and remote sensing even during the 1990s, when Tim Olson, an astrophysicist who had been teaching physics, began teaching remote sensing. The remote sensing course was theory-based because no software was available to the tribal college at that time. In addition, according to Tammie, GPS was commonly used on the Flathead Reservation long before it was commonplace in the rest of the country, with large backpack-style and even back-of-utility-van units being used in the late 1980s and early 1990s.

Tammie Grant, true to her name, began writing grant proposals to the Landsat Data Continuity Mission at NASA to fund a series of workshops at SKC. Each summer starting in 2001, she journeyed from her home in California to teach at the college in Montana. Her workshops evolved from being focused on teachers to being centered on students, where Tammie believed a larger and more long-term positive impact could occur.

Figure 1. Sierra Momburg, former Salish Kootenai College student, in the Sweet Grass Hills, aboriginal territory for the Blackfeet Tribe. Photo by Tammie Grant

She formed a powerful team of geospatial talent that included Volker Mell, a GIS Coordinator, and Cindy Schmidt, a NASA Ames Research Center employee who also shared an interest in teaching geospatial workshops on tribal lands. The summers were a combination of workshops and internships. The "geospatial internships" were seven weeks long in the beginning and provided hands-on experiences in the geospatial lab at SKC. Equally important, though, was spending quality time in the field each day, where the participants analyzed the landscape and collected data on that landscape. Typical mornings would see participants analyzing data through remote sensing and GIS software, including ArcGIS, and during the afternoons, they would disperse to the field to collect data that they would analyze the following morning. Although the formal workshops were student-focused, tribal college instructors and even some primary and secondary teachers were invited to attend. However, Tammie felt that the best workshops were with student-only or teacher-only classes. She noticed in some instances with combined classrooms that some technically savvy students were catching on so quickly to the software that the teachers (from a different generation) tended to lag behind and stay behind. Separating the groups, she felt, led to a less stressful feeling in the classroom. Combining fieldwork and lab work satisfied multiple learning styles because, according to Tammie, "something rang true with them either in the lab or in the field." Local data was always used as the starting point for investigations, and the type of local data would be tailored to the participants no matter whether the workshop focused on the Flathead, Pend d'Oreilles, or Blackfeet.

Rather than see learning coming to an end at the conclusion of each summer institute, the students were encouraged to spend all summer further developing their projects. Some were extended throughout the entire academic year that followed, funded by additional grants sought by Tammie. The tribal government and college received the data, conclusions, and GIS output from these projects and were able to turn them into long-term studies.

As the interest in the summer internships progressed at SKC, NASA's interest in working with Native American students increased, and together Tammie and NASA's Landsat Data Continuity Mission (LDCM) worked to provide summer research experiences for tribal college students. In addition, with the American Indian Higher Education Consortium (AIHEC) (http://www.aihec.org), which provides leadership and influences public policy on issues surrounding American Indian higher education, a larger program was developed and funded for teachers and students called Summer Research Experience (SRE). AIHEC also promotes and strengthens indigenous languages, cultures, communities, and tribal Nations, and through its unique position, serves member institutions and emerging tribal colleges and universities. This was a fruitful partnership that saw incredible student growth through four years of ongoing projects. This growth included the connectedness that students had with their own community and tribal lands and the desire to help their own communities, an interest in using data to solve problems, an interest in using GIS and remote sensing technologies as a career, and an integration of spatial analysis with Native ways of learning and knowing.

Featured projects

One project that the SKC students tackled was a GIS-based study of grizzly bears. Bears have long been an important part of the culture for many Native American Tribes, considered as bringing strength, protection, leadership, and health to Native communities. A group of students worked with tribal scientists to place GPS-enabled collars on grizzly bears and then monitor the locations of the bears using GIS. With GPS, GIS, and remote sensing, the students also mapped and evaluated the health of cottonwood riparian forests, monitoring the change in the spectral signature over the years. Cottonwood forests and other hardwood stands provide the bears with a rich food source and house high protein moths that some bears seek. This project has been in place for over fifteen years, becoming more valuable as time passes. All information is maintained and delivered to the tribal wildlife staff under the supervision of Art Soukkala and SKC Wildlife Instructor Bill Swaney.

Another project created and nurtured by these efforts focused on wildfire management. Fire has impacted the Flathead Valley in many ways, particularly the 2004 fire in nearby Glacier National Park. Tribal leaders and citizens are acutely aware the Mission Range on the reservation has not burned in 100 years. Students and educators used GIS and remote sensing to examine the history of fire in the region and the current risk. Current risk was determined by examining vegetation, precipitation, land use, infrastructure (railroads, roads, and buildings), land ownership, hydrology, and other layers. One factor in determining fire risk is the length of time that snow lies on the ground in the spring, which depends not only on the snowfall during the previous winter, but also on elevation, direction of slope, and degree of slope. From a digital elevation model (DEM), GIS analytical functions could determine the direction and degree of slope, which were combined with the vegetation and land-use layers for a more complete assessment. Again, a field component was important. Students, along with LDCM scientists, examined the impact of recent and historical burns and measured the height of regrowth of specific species. Some students worked with tribal forestry professionals and had the opportunity to drive through high-risk areas on the reservation with fire managers. The students learned about science and job skills in a job-shadowing environment.

Another project that came out of these efforts involved visualizing the landscape, connecting that visualization to cultural sites and places of special significance to the Tribes, and connecting those places to Native languages. A fly-through visualization of a culturally significant trail on the Blackfeet Reservation was created. This visualization followed a trail and told parts of a Blackfeet story. GPS-collected latitude and longitude points from those traveling on the trail on foot and on horseback were used to collect the points that resulted in the trail being accurately mapped and then visualized using 3D GIS tools with Landsat imagery as the base. The visualization was illustrated by spoken dialog in two languages—Blackfeet and English. Tribal Elders provided the stories. It became a teaching tool at Blackfeet Community College and in their K-12 school programs as well as for staff and visitors at Glacier National Park. Another visualization was made of the Flathead Reservation showing certain cultural sites and natural resources significant to the Salish, Kootenai, and Pend'Oreille People. This provided a

Figure 2. Sierra Momburg, former Salish Kootenai College student on the Blackfeet Indian Reservation. Chief Mountain rises in the background. Browning, Montana. Photo by Tammie Grant

unique opportunity for a wide variety of Native and nonnative people to hear the cultural stories with a landscape visualization. As maps have done for thousands of years, the visualization helped people understand the geographic, historical, and cultural significance of these cultural lands. Even though a year of hard work was required to incorporate the Native language hyperlinks and stories into the visualization, many of these stories had not ever been heard in many of the classrooms. Particularly important were the stories being offered in the Native languages, which according to Tammie, are "on the edge of extinction."

Mentoring was also a part of this learning so that the summer workshops would not be seen as solitary incidents. Tammie sought previous students as graduate mentors for the workshops as a way to build a sustainable program. Not only did the program grow, but also the content knowledge, GIS skills, and most importantly, the confidence of each student as they helped others. Don Sam and Mark Couture were two among the many students that were trained in geospatial technologies and served as mentors for some of the workshops. Tammie has tracked students who have moved through the program, and many of them are in environmental or land management careers with their Tribes, using GIS on the job as a daily decision-making tool.

Conclusion

Tammie's current projects are diverse and include other indigenous Peoples from overseas. She examines how Native cultures perceive and map their own lands. Through kids' eyes, Tammie is interested in capturing how tribal youth perceive their land through family stories and other cultural influences. She is linking indigenous students in Russia with US Tribes. One of her dreams is to set up an international multicultural science camp for kids. She is interested in mapping things that are not traditionally mapped within a GIS, such as stories and song lines. She is forever concerned about asking the right questions and using the right data—with GIS-based projects and beyond. Her thesis is "if we are not asking the right questions, then we won't gather the right data and most effectively address the key issues that need to be addressed." Despite the diversity of these projects, underlying each of them is a deep connection to the spatial perspective, to learning, and to place.

Building and sustaining a GIS education program for Indian Country at Haskell Indian Nations University

Carol Bowen, John Kostelnick, Dave McDermott, Danny Rowland, and R. J. Rowley
Haskell Indian Nations University
Lawrence, Kansas

Education for Indian Country through Haskell Indian Nations University

For almost twenty years, Haskell Indian Nations University has included GIS in its curriculum. The intent of the program has always been twofold: to equip students with skills they can use in their careers, particularly in service to indigenous communities, and to encourage study in science and mathematics. These efforts have served the broader mission of Haskell, a four-year university in Lawrence, Kansas. Haskell is one of only two tribal universities administered by the Bureau of Indian Affairs (BIA). It is a center of Native American higher education, serving Tribes throughout the United States, and holds great symbolic value for the Native American educational community. In addition to the traditional university missions of teaching and research, Haskell commits itself to the cultural preservation of indigenous communities. One method by which it furthers that goal is instruction in the techniques of mapmaking, equipping graduates with the tools with which they can help their communities take charge of the environmental management and cultural protection of their lands.

GIS education at Haskell

Over the decades, Haskell has collaborated with federal agencies and academic institutions to build a GIS program. In the 1990s, the program used Idrisi software and staff from the mathematics department and the local US Geological Survey (USGS) office. That effort ended when funding from USGS and institutional support disappeared. The GIS program returned when Haskell became a partner in the National Science Foundation's Polar Radar for Ice Sheet Measurement (PRISM) program and continued with the Center for Remote Sensing of Ice Sheets (CReSIS). In both of these programs, Haskell has been responsible for providing mapping and GIS support for polar research.

The geography program at Haskell has always incorporated undergraduate research projects with traditional coursework. The projects fall into three categories: service to the Haskell campus, research on environmental and social issues on tribal lands, and, through our partnership with the CReSIS, polar research. This broad range of projects allows students to focus on areas of personal interest. In the case of GIS projects for the Haskell campus and projects focused on tribal lands, they also allow students to use their growing talents as geographers and GIS scientists to give back to their communities. The polar science projects allow students to reach well beyond their personal experiences and take part in cutting-edge research on ice sheets and impacts of climate change.

Examples of GIS-based projects

In the last five years, Haskell students have worked on GIS projects as far afield as the bedrock under Antarctica's ice and as nearby as the walking paths of the Haskell campus. Projects explored cultural stories (improving maps of the Trail Where They Cried), business stories (exploring potential for in-stream hydroelectric generation near tribal lands), and science stories (modeling impact of sea level change). Students who work on these projects are recruited from the intermediate and advanced GIS classes and are paid a modest hourly wage. They typically work about ten hours a week during the school year and twenty hours a week during the summer. Examples of recent projects include mapping medicinal plants on the Winnebago Reservation, mapping accessibility on the Haskell campus, and modeling subglacial bedrock in Greenland and Antarctica.

Mapping plants on the Winnebago Reservation

The Haskell GIS lab was approached by staff from the Winnebago Reservation in Nebraska. Working with their own students and an ethnobotanist from the University of Kansas, the college had assembled native plant census data for the reservation but did not have the resources to turn that data into an atlas. A Haskell student, Danny Rowland (Cherokee), took on the task of plotting the raw data. Initially, we

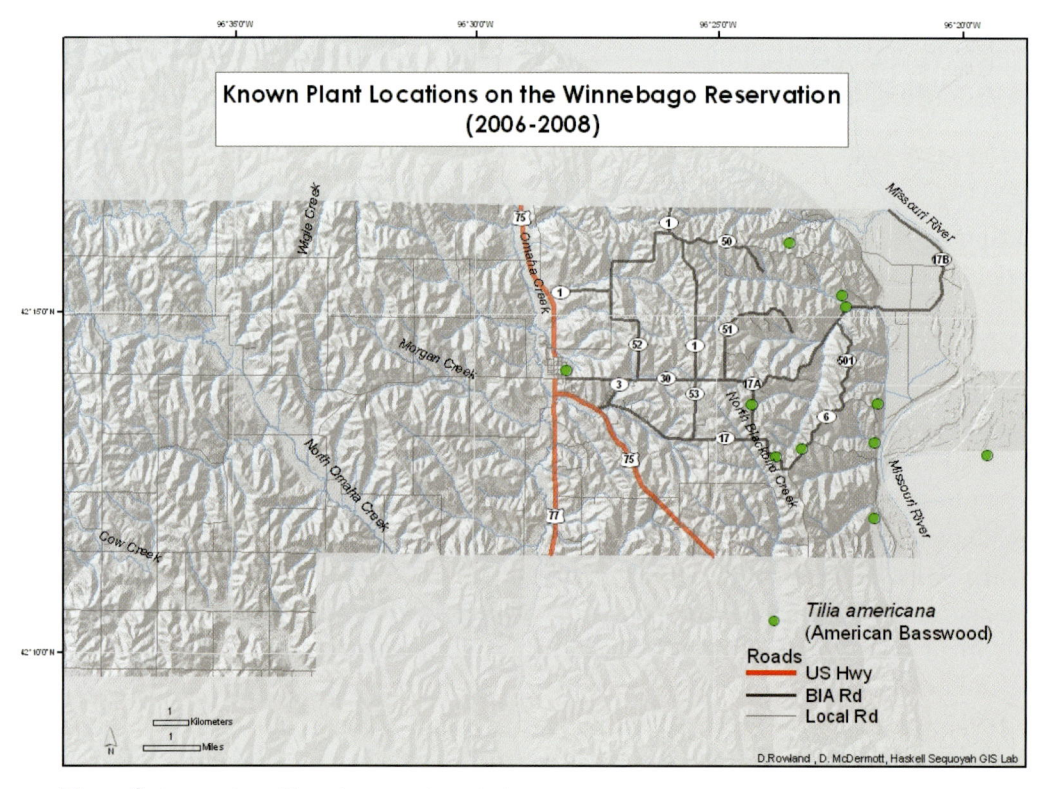

Figure 3. A page from Winnebago medicinal plants atlas. Courtesy of Haskell Indian Nations University

anticipated producing a single poster-sized map. However, many species of interest were often found so near each other that data for any individual species would have been lost on a single map at a coarse scale. Danny therefore decided to make a map book with separate pages for each species at a larger, finer scale.

This project was more than a service to our colleagues at Little Priest Tribal College in Winnebago, Nebraska. It provided an opportunity for a Haskell environmental science student to work with real data and to discover some of its limitations. It also provided an opportunity to collaborate with an ethnobotanist who may turn out to be Danny's advisor when he begins his graduate study in geography.

Mapping accessibility on the Haskell campus

Over forty years ago, the Architectural Barriers Act committed the federal government to design and build facilities that are accessible to persons with disabilities. Interviews with university staff, however, revealed that there was no map indicating access ramps, elevators, and curb cuts that are necessary for someone with limited mobility to navigate the campus. A Haskell student, Marshall Bass (Winnebago), took on the task of finding and mapping every accessibility feature on the campus.

Data collection was a major part of the project. Marshall had to walk the campus, noting the location of accessibility features. He also had to measure the effectiveness of access features, taking note of design failures such as curb cuts that lead to dangerously steep ramps and sections of buildings that were inaccessible due to internal staircases. The project also required the creation of new map symbols, going beyond the generic wheelchair graphic, to indicate specific features such as curb cuts, ramps, and paths of varying steepness.

Figure 4. Detail from the Haskell campus accessibility map showing new symbology.
Courtesy of Haskell Indian Nations University

Modeling impacts of global climate change

Understanding the potential impact of global climate change on sea levels around the world turns on an understanding of the mass of ice over Greenland and Antarctica and the rates at which it may be changing. Sea level rise, in turn, will have consequences for indigenous communities throughout the world and for coastal communities generally.

For several years, Haskell students have been involved in measuring and mapping the surface of bedrock under the glaciers of Antarctica and Greenland. Those projects include modeling outlet glaciers—the fast-flowing ice streams that funnel ice and water from the interior of the continent to the coast—and mapping under-ice terrain to better understand the geologic history.

Richard LaBrie (Cherokee) recently worked on a project to map the surface of the Gamburtsev Mountains, buried under a thousand meters or more of ice in Antarctica. The work involved interpolating ice thickness from point data collected by airborne ice-penetrating radar. Once reliable thickness interpolations were created, those values were subtracted from a digital elevation model (DEM) of the ice surface to yield an image of a mountain range no human has ever seen. Richard and his colleagues travelled to San Francisco to present their results at the annual meeting of the American Geophysical Union.

Figure 5. A 3D representation of the Gamburtsev Mountains, Antarctica, buried under ice. Courtesy of Haskell Indian Nations University

Sustaining a program

The experiences at Haskell provide several lessons for other small institutions serving Indian Country. First, project-based learning works particularly well for GIS education and provides a mechanism by which a college or university can avoid the pitfall of allowing its curriculum to become purely vocational training while at the same time providing students with very marketable skills in geography and GIS. Second, just as collaboration is essential in doing science, it is essential in teaching science. Collaboration with other universities is one way to keep both the program and the faculty in a small university connected with new directions in research, and even provide an opportunity for the smaller university to contribute to a broader research project. Finally, the recognition that even well-designed and respected programs at small institutions are vulnerable to great damage from relatively small changes in staff, funding, or institutional practices.

The project approach to GIS education described here supports the role of GIS as a legitimate field of university study. By showing how GIS can be science, as well as support science, the projects at Haskell open a path for students who want to specialize in GIS in their careers or as graduate students. For students who do not anticipate careers in geography or geographic information science, the same projects allow them to build a set of GIS skills that they can apply as practitioners in their chosen fields. Students report that their work in the GIS lab has prepared them for careers by giving them the chance to do meaningful work in an environment in which their judgment is respected.

Collaboration with other institutions, particularly the large research universities that partner in the CReSIS program, has provided much more than financial support to the Haskell program. The partnership has provided an academic path for students who want to pursue graduate study; those students can make a relatively seamless transition from working as a Haskell undergraduate research assistant to an appointment as a University of Kansas (KU) graduate research assistant. Particularly for students who are the first in their families to pursue advanced degrees, any mechanism to smooth the transition to graduate school increases the likelihood that the student will be successful. The partnership also provides Haskell students with access to leading scholars. Senior faculty from CReSIS and KU routinely guest lecture in the Haskell GIS program. Finally, the partnership allows the GIS faculty at Haskell to stay in touch with new scholarship in geography and geographic information science. The challenge of keeping faculty up to date in their fields while working at teaching institutions is well recognized; it is particularly acute in fields such as GIS, where the technology and its applications change dramatically from year to year.

These necessary collaborations go beyond association with other universities. Haskell routinely calls on guest lecturers from the vibrant engineering community in Kansas City. Colleagues at Wilson and Company, an engineering firm with a large air mapping operation, and Black & Veatch, a diversified engineering firm, share their experiences with students and faculty. Faculty at Haskell have collaborated with staff at Wilson and Company on conference presentations, further strengthening the connections between the campus and industry. There is also a dependence on the school's association with Esri, whose staff and publications keep the faculty informed of developments in the broader profession.

Even well-designed programs, however, are vulnerable to institutional changes, particularly when they are funded exclusively by grants. Sustainability of the GIS program is a constant struggle at Haskell and may be a similar challenge at other small institutions, including tribal colleges and universities. Although grants are useful for building GIS programs, stable funding is essential to sustain GIS programs over the long term.

Community mapping

Allyson Kelley, Master in Public Health, Certified Health Education Specialist
Ed Aucker, Big Horn County Emergency Planner
Avis Spencer, Fort Belknap Tribal Health
Eric Wood, PhD, USGS Earth Resources Observation and Science (EROS) Center

Introduction

Community mapping through the Montana Indian Country CARE Project (MICCP) was created through a partnership with Rocky Mountain College (RMC) and USGS Earth Resources Observation and Science (EROS) Center. This partnership combined academics, tribal colleges, community needs, and mapping expertise to form the first-ever series of community mapping workshops on these reservations.

Background and objectives

Four separate reservation communities participated in this project: Fort Peck, Fort Belknap, Northern Cheyenne, and Crow. With funds from the Environmental Protection Agency (EPA) CARE Project, RMC, and US Geological Survey (USGS), a series of community mapping workshops were held at tribal colleges. These workshops used computer labs set aside for the workshop and the generous support of tribal college staff.

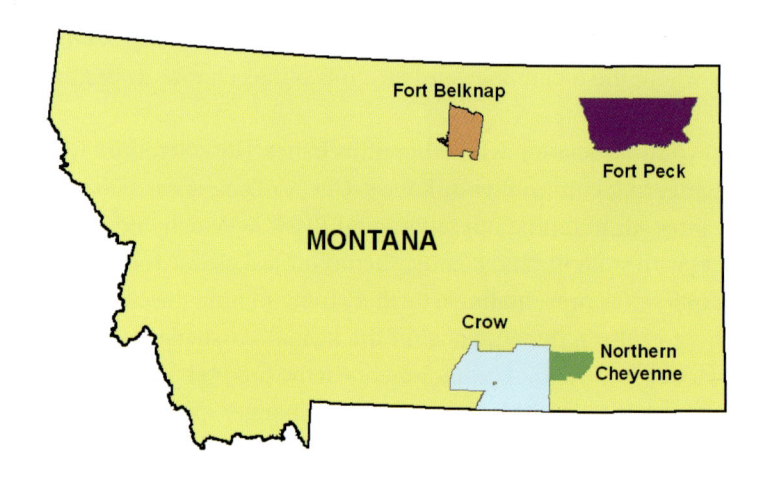

Figure 6. Community mapping workshop locations in Montana. Courtesy of Rocky Mountain College, Big Horn County Emergency Planner, Fort Belknap Tribal Health, USGS Center for Earth Resources Observation and Science

The MICCP was a community-based initiative to reduce toxins in Indian Country. Birthed out of lasting relationships with tribal colleges, RMC secured federal funding through the EPA, Level II Grantee funding, for a two-year period, 2006–2008. Tribal colleges donated computer labs and classroom facilities, Esri provided the GIS software, and the USGS provided the instruction and technical support. Tribal GIS professionals were recruited to assist in the classroom.

The objective of the MICCP was to create healthier communities by hiring tribal members to coordinate community activities that focus on exposure to toxic substances in the air, water, and soil, and how to mitigate such exposures through educational interventions, community clean-ups, and targeted site remediation projects.

Although the MICCP focused on toxic substance reduction, the community mapping workshop attracted participation from tribal environmental managers, land professionals, students, teachers, and tribal council members. By attending the workshop, participants enhanced their capacity for using, understanding, and creating maps in their respective communities. Participants also contributed to community toxic reduction efforts and environmental initiatives created from MICCP. Mapping flyers were sent to all tribal agencies, professionals, college students, and teachers. No mapping experience was required to attend the workshop.

The first day of the workshop included the creation of hand-drawn reservation maps, slide presentations, field activities on handheld Garmin GPS receivers, a trial activity of mapping, and computer applications.

After introductions, students were asked to draw maps of their respective reservations. This exercise provided insight into the participants knowledge of maps (items such as scale, legend, size, coordinates), and allowed students to get to know each other.

Educational research

Did these workshops effectively teach GIS skills and knowledge? Anyone who has trained or taught has confronted this question. At Fort Belknap, Chief Dull Knife, and Little Big Horn Colleges, students were given a pretest and posttest to measure their knowledge of community mapping and GPS/GIS. Some participants attended only part of the workshop. The scores reflect whether someone attended the complete workshop or only partially. The test comprised ten questions that related to basic definitions used in mapping:

1. What is community mapping?
2. What does GPS stand for? What does it do?
3. What does GIS stand for? What does it do?
4. What is a map legend?
5. What is a map scale?
6. What is a map projection?
7. What does 1:24,000 mean on a map?
8. What two key pieces of information do you need to enter into your GPS unit to use it accurately with a hard-copy map?

9. What does UTM stand for?
10. What is a topographic map?

The scores improved in all cases after the workshop, reflecting the knowledge gained from the workshop. In figure 7, "N" represents the number of people who took the test at Fort Belknap, Little Big Horn, and Chief Dull Knife Colleges. The mean test score improved by 2.24 points from the pretest to the posttest for students attending the full workshop and by 2.26 for students attending only one of the two days of the workshop. Test scores for participants attending the full workshop were slightly higher than those attending partial days. Students who attended the second day could have had previous work experience and knowledge of GPS/GIS. The second day included hands-on activities with GPS/GIS that required the student to demonstrate competence.

Community Mapping Workshop		Pretest score	Post-Test score
Participants attending days 1 & 2	Mean	5.17	7.41
	N	22	22
	Std. Deviation	2.48	2.61
Participants attending only part of days 1 & 2	Mean	4.9	7.16
	N	34	34
	Std. Deviation	2.65	2.61

Figure 7. Pretest and posttest data for the community mapping workshop.
Courtesy of Rocky Mountain College, Big Horn County Emergency Planner, Fort Belknap Tribal Health, USGS Center for Earth Resources Observation and Science

Workshop goals

The primary goals of the workshop were to (1) introduce CARE community members to GPS/GIS while increasing the awareness of environmental toxicants and/or stressors, and (2) attract more community members to become engaged and active in CARE toxic reduction activities. Secondary goals were to determine the interest and need for mapping toxic sites in Indian Country, assess the interest and need for further education on the use of GIS and GPS, and to develop a community mapping work group for future mapping projects.

Goals of the MICCP mapping workshop:
1. Explain what community mapping is and how it can be used to address specific community issues.
2. Outline a plan that informs local residents about community mapping and encourages their participation.

3. Demonstrate the basic principles of map reading, including topographic maps.
4. Collect and consolidate data pertinent to community mapping and add to existing USGS topographic maps.
5. Use a handheld GPS unit and ArcGIS Desktop software to document identified community environmental problems.

It quickly became obvious that community members wanted to use GIS in different ways. Examples included the desire to map open dump sites, high fatality car crash sites, cultural preservation areas, recreational fishing locations, cemeteries, road signs, speed bumps, areas where toxic chemicals may be stored, burned buildings, old mining sites, and historical areas.

The workshop was designed assuming participants had no GIS experience. Participants who attended with experience still learned something new and found the workshop helpful, based on comparing pretest and posttest results. ArcGIS Desktop software was used in the workshop. We found that the experience of GIS users in the community to be varied. People had different ideas about GIS and using GPS. These ideas shaped their experiences in the classroom. For example, a fire chief was interested in mapping toxic and flammable liquids. The Bureau of Indian Affairs (BIA) leasing representative was most interested in land leases, boundaries, and maps. Environmental staffers were generally interested in open dump sites or watersheds. These differences, in combination with varying experience, contributed to a unique GIS experience for everyone involved.

The most challenging aspect of GIS was the use of computers for individuals who were not familiar with this technology. There were several tribal GIS representatives who were recruited to attend the workshop and assist the instructors. These representatives assisted with computer challenges.

There were a number of maps created from this project. At Fort Peck, community members mapped burned buildings; this information when combined with public health information solved a complex public health problem (see the section in chapter 5 on using GIS to analyze asthma on Fort Peck lands in Montana). In another case, a map of motor vehicle crashes (MVCs) was created by collecting waypoints where crashes occurred. This information was cross-referenced with a paper map with multicolored pens where MVCs were recorded. A tribal cultural committee at the Northern Cheyenne Reservation used ArcGIS to map a historical and cultural site for preservation.

The final outcome of this project was increased awareness of GIS as it relates to community and the environment. Another benefit was improved communication between different tribal groups using GIS. Finally, this project articulated the need for additional GIS training and support through the submission of two grant proposals.

A total of seventy-two community members signed up for the training at four reservations in Montana. Of these, forty-three people attended the training. Twelve participants missed at least one day of the two-day workshop. Tribal environmental departments, BIA employees, and tribal health workers made up the majority of student participation.

Workshops were held June 10–20, 2007, starting at Fort Belknap College in Harlem, Montana. Training continued June 13–14 at Fort Peck Community College in Wolf Point, Montana. Instructors continued their journey south to Chief Dull Knife College in Lame Deer, Montana, on June 15 and 18. Workshops were completed on June 19 and 20 at Little Big Horn College, Crow Agency, Montana.

Workshop content

The workshop curriculum was developed by a senior scientist at USGS EROS. The content was based on feedback from Allyson Kelley, Project Director of MICCP, and community meetings where tribal members voiced their concerns about toxic sites, pollution, and energy development. Reference books, ArcGIS Desktop, and Garmin GPS receivers were given to participating tribal colleges upon completion of the workshops. The following reference books were used:

- *Community Mapping Handbook, Mapping Our Land: A Guide to Making Maps of Our Communities and Traditional Lands.* Alix Flavelle. ISBN 1-55105-376-4.
- *Making Community Connections, the Orton Family Foundation Community Mapping Program.* Connie L. Knapp. ISBN 1-58948-071-6.
- *Community Geography, GIS in Action, Teacher's Guide,* Lyn Malone, Anita M Palmer, and Christine L Voigt. ISBN 1-58948-051-1.

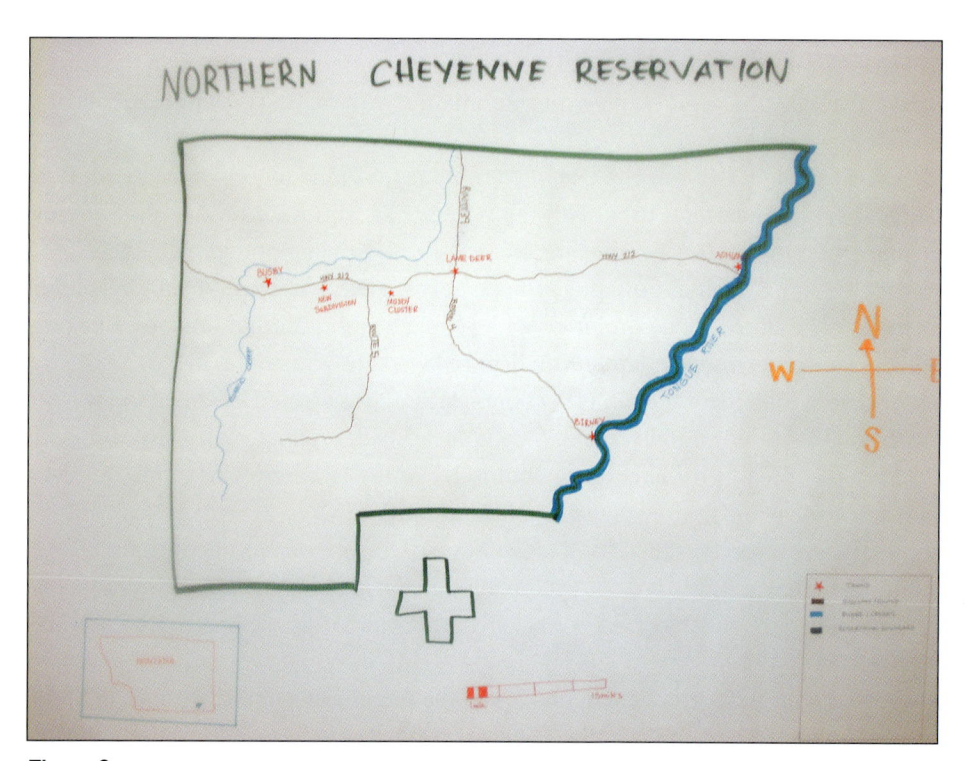

Figure 8. In this activity, students were asked to draw a map of the Northern Cheyenne Reservation before learning anything about maps. Courtesy of Rocky Mountain College, Big Horn County Emergency Planner, Fort Belknap Tribal Health, USGS Center for Earth Resources Observation and Science

Each workshop began with an overview of community mapping. Students were asked to introduce themselves, their affiliation with their Tribe, and their reason for attending the workshop. Students were then asked to define community mapping. Answers were varied and included in the pretest and posttest analysis. After this introduction, students were asked to draw maps of their reservations. Participants used markers, rulers, and poster-size paper to draw maps. Students affixed them to walls and explained the boundaries and topics of each map.

Figure 9. A workshop begins with discussion on the goals for community mapping; here, at Chief Dull Knife College on the Northern Cheyenne Reservation. Courtesy of Rocky Mountain College, Big Horn County Emergency Planner, Fort Belknap Tribal Health, USGS Center for Earth Resources Observation and Science

The introduction to community mapping was followed by a series of slide presentations to share community mapping definitions, examples of community mapping, and the benefits to a community. Presentations and worksheets reinforced learning objectives and provided the framework for independent field activities. Presentations provided students with examples of how and why communities utilize GPS/GIS mapping. A community in Africa mapped coffee crops to determine geographic locations of superior coffee crops. Indigenous Peoples in Canada used mapping to establish ancestral boundaries. Communities use mapping to determine pollution sources, dump sites, industrial areas, and housing structures.

After several hours of classroom presentations, students ventured to an outside location and mapped parking lots, city blocks, alleys, and college areas. Students entered waypoints and set a track log with assistance from the instructors. Students returned to the classroom and transformed the waypoints and tracks into ArcGIS shapefiles. This activity concluded day one of the workshop.

Figure 10. Fieldwork is an important component of community mapping workshops, including this one at Fort Peck Community College. Courtesy of Rocky Mountain College, Big Horn County Emergency Planner, Fort Belknap Tribal Health, USGS Center for Earth Resources Observation and Science

On day two, after a brief introduction to community mapping, students were asked what they wanted to map. Student lists included animal feedlot operations, noxious weeds, open dump sites, cemeteries, chemical storage areas, burned down houses, vacant houses, streets, cultural sites, and water sampling areas. Instructors assisted the class in developing day-two activities. Students paired up, driving to identify areas to map selected subjects.

After the field activity, students returned to class with GPS units and field notes in hand and a better understanding of community mapping. Students first downloaded waypoints and tracked logs in a file format. Students queried, inserted, and printed maps with the data collected during the field activities on day two.

The instructor's impression of field activities and community needs are as follows:
1. The need for technical expertise and mapping Montana reservations is overwhelming to tribal professionals and community members who want to produce maps.
2. Basic mapping of houses, street names, cemeteries, community areas, water supplies, utilities, hazardous waste sites, and land boundaries is currently inadequate on several reservations.
3. Students came to the workshop wanting more technical expertise and time to map items than their current jobs allowed. For example, a BIA professional wanted to map land boundaries of allottees and leases. This project required a high level of technical expertise, equipment, software,

and time, and certainly more than could be attempted in a two-day workshop. While the community did not map this, the workshop reinforced learning objectives previously taught to the student.
4. Additional expertise and instructors to accompany students on the second day's field activities would be helpful.

Recommendations for future workshops are as follows. Future workshops should include more time for ArcGIS and field activities. Additional instructors would be helpful to accompany students during data collection in the field. There should be a greater involvement from tribal agencies that have mapping needs. The development of mapping subjects should be done prior to the workshop. Helpful input would be from a query to tribal agencies before the workshop. Also helpful would be involvement from all tribal professionals working in GIS and GPS. This would allow relationships to be created and minimize the duplication of efforts associated with mapping tribal communities.

Reaction from workshop participants

Selected responses from workshop participants are provided below. The following e-mail was from Ed Aucker, the Big Horn County Emergency Planner. He attended one of our community mapping workshops and continues to use GIS at Crow and Northern Cheyenne.

When I took the class I was curious about the GIS concept and wanted some introduction. The class did that and gave me enough direction that I sought out additional training and mentors. It has been perhaps 4 years now and I am still at what I would call the "representational" stage of GIS. By that I mean, I can gather data, and represent what is there, but I have not yet progressed to the synthesis of new knowledge that I know GIS can provide. But I know people who can do it . . . .

Just yesterday I was in Lame Deer working with them on their efforts to map the area well enough for E911 . . . .

Share the data. Too often bits of data are held too closely either out of corporate policy or a power thing. But held data is lost data—at least lost to the community. I was in Crow—Susette needed information about prairie dog towns. She had been provided a paper map without enough data on it to be of any value so she was trying to figure out how to (re)!! map the towns—her time would be better spent figuring out how to get access to the existing shape files which were used to produce the worthless paper maps. With those files in hand I could build her a set of new attributes that would meet her needs for the grant she is working on—what other critters live there? Is the town shrinking or expanding? Is there public access to the town? Instead either because of policy or jealousy, Susette will wind up spending the summer remapping what has already been mapped.

Ed

The write-up below came from Avis Spencer, Fort Belknap Tribal Health Injury Prevention Assistant, who attended one of the community mapping workshops:

The benefits of community mapping are endless, in my position at tribal health mapping could be utilized in many ways. Water wells (dry, abandoned), septic tanks and other public utility lines could be mapped, as well as motor vehicle crash sites, clusters of illness, cultural sites and the reality of rural addressing, which we still do not have completed to my knowledge. Schools, clinics, churches, tribal buildings, emergency shelters, etc. Every program could benefit from community mapping, forestry, land department, utilities, transportation, roads, planning, health, housing, social services, EMS, law enforcement, volunteer fire department, etc. This year my goal is to actually find the time to attend some free Esri training @ the BIA facility in Albuquerque and map some of the above.

Avis Spencer—Fort Belknap

Summary

The tribal government leadership benefits from these community mapping efforts in a variety of ways, primarily from an increased number of GIS users in the community and increased knowledge of environmental impacts in those communities.

Tribal colleges and governments received copies of ArcGIS Desktop from Esri, and community mapping books were donated to the library for checkout.

Through this project, community members and scientists alike experienced the process of making a map to communicate a story to impact change. Unforeseen benefits of workshops included an increased communication among all Tribes involved, offering a second workshop with Northern Cheyenne and Crow in November 2007 (due to the high demand for more GIS training), and a remote sensing workshop offered to community members from the four reservations.

The grant that funded this project has ended. However, valuable relationships are now in place for future mapping classes and more training with USGS and Esri.

8

Photo by Joseph Kerski

Fostering sustainable tribal agriculture and rangelands

Traditionally, Native American communities have honored and relied upon the balance of humans, sun, water, soil, air, vegetation, and wildlife for a healthy and productive community. As such, Native Americans can be considered the first farmers and ranchers of the Americas. Native methods of growing and cultivating food and meat continue to influence the agriculture industry today. Their traditional emphasis on sustainable practices for producing, distributing, and consuming food continues to play a significant role in the well-being of tribal cultures and communities. This chapter explores how GIS helps tribal communities increase agricultural production, reduce costs, and manage land more efficiently while maintaining the culture and traditions of the Tribe. The ability of GIS to analyze and visualize agricultural environments and workflows is the key to its many benefits to the agriculture industry. Pest control, cattle grazing management, precision farming, and water resource management are but a few of the areas where GIS is being used to improve agricultural business practices. Producers can use GIS to better manage their farms by creating reports and maps that give them a unique perspective on their operations. The powerful analytical capabilities of GIS offer an array of options for visualizing farming and ranching conditions, as well as measuring and monitoring the effects of management practices. Central to most agriculture programs these days is growth with sustainability, improving production while reducing ecological impacts. The stories in this chapter explain how several Tribes are using GIS to accomplish just that. On the Chilicote Ranch of the Ysleta del Sur Pueblo in Texas, Beau Barela describes how GIS and remote sensing were used to create a land classification system, which in turn was used to manage wildlife and prevent cattle overgrazing. On the Flathead Reservation in Montana, Mickey Fisher explains how GIS was effectively used to increase agricultural yields through controlling the grasshopper population.

Using GIS for ranch management

Beau Barela, GIS Specialist, Ysleta del Sur Pueblo
El Paso, Texas

Chilicote Ranch is Ysleta del Sur Pueblo's largest property, spanning 70,359.9 acres, located within the counties of Jeff Davis and Presidio in West Texas. Chilicote Ranch was purchased in 1998. Since then the ranch has been used for many ventures; however, the main business enterprise of Chilicote Ranch is cattle ranching. Chilicote Ranch's manager and staff have a wealth of information and expertise on the day-to-day operations and the land itself. However, the GIS Division was needed to illustrate the terrain of the ranch and to identify locations of water storage tanks, pastures, fence lines, and pipelines. This illustration, through the use of maps created using ArcGIS, is important for communication between different tribal agencies and the ranch management team. Maps displaying the locations of resources and terrain features allow for more sustainable planning through either interagency work or outside agency tasks such as working with contractors, insurance agents, or adjoining ranchers.

In addition to basic resource and terrain identification, further land classification was needed to accurately assess vegetation. This additional analysis is used for the purpose of locating and relocating cattle into prime grasslands to prevent overgrazing. Analysis of Chilicote Ranch vegetation was completed using remote sensing software. This software allowed the GIS Division to determine the varying vegetation types. After a preliminary analysis was complete, ArcGIS software was employed for the generation of a vegetation classification map.

This vegetation classification is used to help track wildlife, such as deer, antelope, and Aoudad. As an aside, the Aoudad sheep were brought to Texas in the late 1940s and early 1950s as a result of World World II soldiers stationed in North Africa who saw how hardy and intelligent they were. Habitat preferences of wildlife are illustrated using data from the classification analysis. Pastures used for cattle grazing are carefully managed using vegetation classification maps, so as not to affect wildlife habitat by cattle grazing. These maps and further analysis are also used for restoration of habitats that have been left in poor ecological condition because of past overgrazing. The GIS Division has aided the newly formed Wildlife Management Committee, which serves to ethically manage wildlife populations and create favorable habitat conditions to sustain wildlife populations. The committee's management efforts are coordinated with the Texas Parks & Wildlife Department, and through these collaborative efforts the Pueblo conservatively allows the harvesting of a set amount of animals prior to the hunting season. By comparing animal inventory data that the Pueblo has gathered and inventory records from the state, the Wildlife Committee determines, for instance, how many mule deer can be harvested for the upcoming hunting season. Exotic animals like Aoudads are treated as management hunts. No caps are placed on

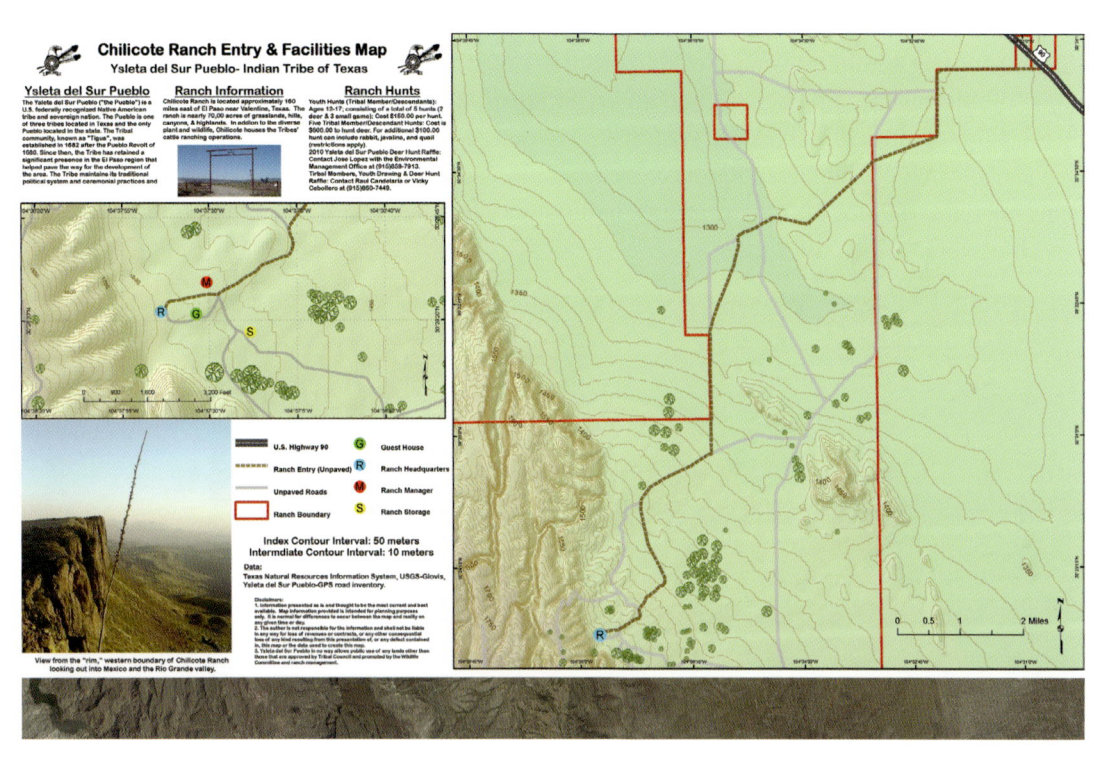

Figure 1. Chilicote Ranch Entry & Facilities Map. Courtesy of Ysleta del Sur Pueblo

these hunts, and their season is regulated by the state. The preliminary and future vegetation classification will assist Ysleta del Sur Pueblo's Wildlife Committee's hunting operations, as properly managed wildlife habitats will result in more wildlife being harvested in a sustainable manner.

The GIS Division has also completed an inventory of deer and antelope populations in order to assist the Wildlife Committee with a significant breakdown of these populations. The GIS Division completed maps with the help of GPS equipment outfitted with ArcPad software. The GIS Division also made use of GPS-enabled cameras for data collection. Importing all collected data into ArcGIS, the GIS Division was able to generate maps that demonstrated density of deer population by pasture and summarized approximate ages of male deer by antler size. Female and fawn populations were also recorded; however, they will not be hunted. Their numbers will be used instead to determine future populations and to verify survival rates of faunal populations. The committee has used this inventory to accurately provide a feasible number of animals that could be harvested for the 2009 hunting season. Future inventories will be completed on a bi-annual basis to keep abreast of any changes in populations of wildlife. After the hunts are completed, the Wildlife Management Committee guides send their records of hunt locations, wildlife type, age of wildlife, and hunter information to the GIS Division. The GIS Division then takes this information, records it in the appropriate geodatabase, and creates maps illustrating the statistics listed above. Being able to provide all of these services in-house has allowed Chilicote Ranch and the Wildlife Committee to be more cognizant of wildlife populations and has allowed the Ysleta del Sur Pueblo to exercise its sovereignty and control of properties using a sustainable approach. In the past, the Pueblo had contracted hunting outfitters to conduct hunts with a flat fee being charged to the hunting outfitters. This became troublesome to the Pueblo since the outfitters never provided a number of animals harvested or the location of animals harvested. With the development of the GIS Division, the Pueblo has been able to actively manage Chilicote Ranch cattle operations and ensure that the appropriate numbers of animals are harvested for hunting ventures, habitat for wildlife is maintained, and future generations will be able to enjoy Chilicote Ranch.

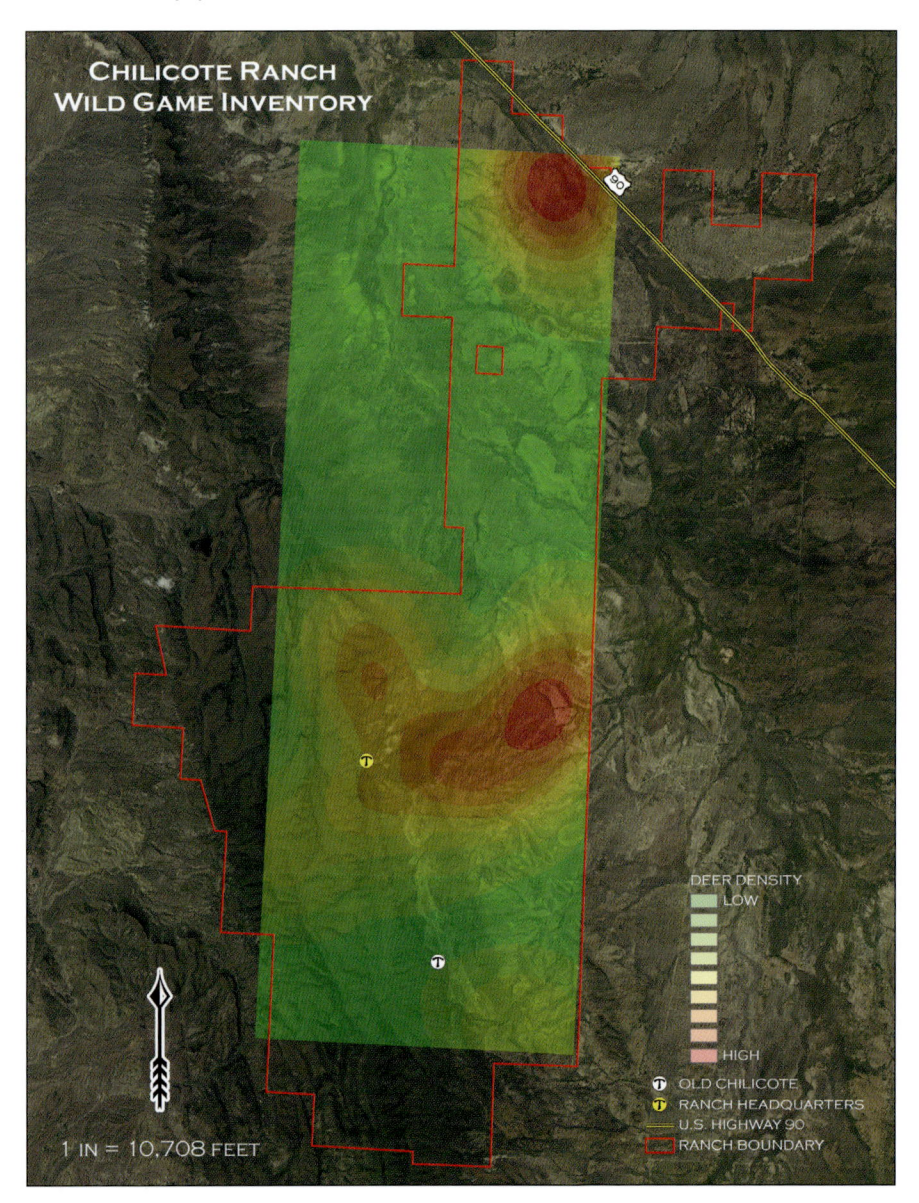

Figure 2. Chilicote Ranch wild game inventory. Courtesy of Ysleta del Sur Pueblo

Confederated Salish and Kootenai Tribes of the Flathead Nation Grasshopper Mitigation Project

Mickey Fisher, GIS analyst
Confederated Salish and Kootenai Tribes Natural Resources Department; Confederated Salish and Kootenai Tribes Environmental Protection Agency/Pesticides Program; Confederated Salish and Kootenai Tribes Natural Resources Department Wildlife Program; Tribal Lands Department, Weeds Management
Pablo, Montana

The Flathead Indian Reservation is located in Northwest Montana and home to three Tribes, the Bitterroot Salish, Upper Pend d'Oreille, and the Kootenai. Confederated Salish refers to both the Salish and Pend d'Oreille Tribes. The territories of these three Tribes covered all of Western Montana and extended into parts of Idaho, British Columbia, and Wyoming. The Hellgate Treaty of 1855 established the Flathead Reservation, but over half a million acres passed out of tribal ownership during the land allotment that began in 1904.

The subsistence patterns of the Confederated Salish and Kootenai tribal People developed over generations of observation, experimentation, and spiritual interaction with the natural world, creating a body of knowledge about the environment closely tied to seasons, locations, and biology. This way of life was suffused with rich oral history and a spiritual tradition in which people respected the animals, plants, and other elements of the natural environment. By learning from the Elders and teaching the children, tribal members can continue their ways of life to this day.

The Confederated Salish and Kootenai Tribes (CSKT) began using GIS in 1988, and the CSKT GIS Program was established in 1990. Overall, GIS has assisted CSKT in analyzing and depicting geospatial data by faster and more efficient means than before GIS was used by CSKT. The Tribe currently uses GIS to serve the tribal population as well as the general public that lives on the Flathead Reservation in Western Montana. In this grasshopper project, GIS helped map the desirable areas for spraying to control the spread of grasshoppers, allowing the budget for spraying to be available to treat more acres, as well as allowing for a better yield for agriculture. Ranchers and farmers have always had to deal with the problem of grasshoppers in Montana. Grasshopper population density depends on a variety of factors, including weather conditions and the availability of food during the different stages of a grasshopper's development. Over the past several years, Montanans have seen an especially dense grasshopper population in specific parts of the state.

The CSKT GIS Program is led by the GIS Program manager and includes a GIS analyst and a GIS cartographer. The majority of CSKT GIS production is aimed toward the CSKT Natural Resources Department (NRD), though all other tribal programs benefit from CSKT GIS throughout the reservation, such as Cultural Preservation, Tribal Education, Tribal Health, and many other programs. CSKT GIS

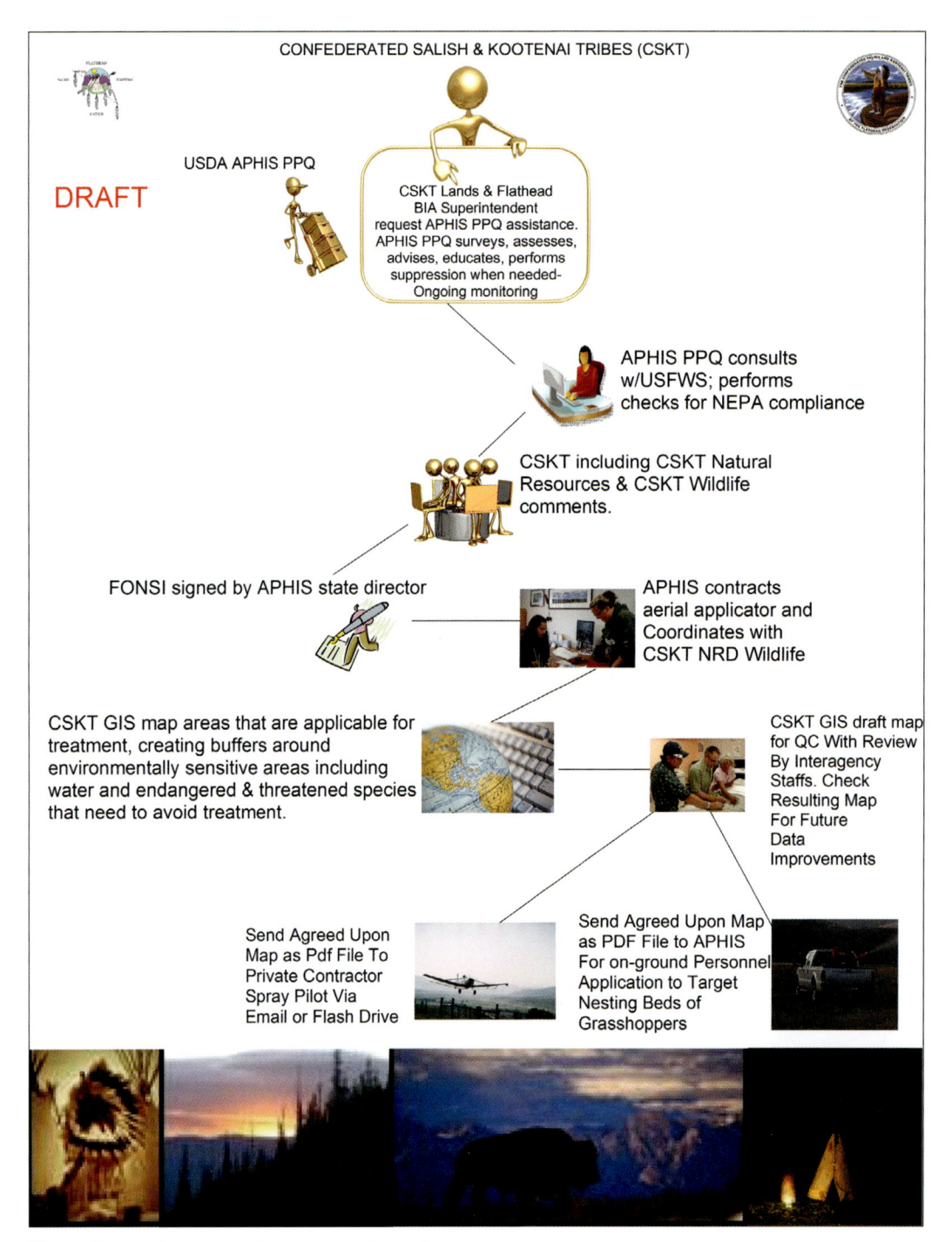

Figure 3. Workflow for the Grasshopper Spray Project. Courtesy of Confederated Salish & Kootenai Tribes

also works with city, state, federal, and other local governments and agencies such as the USDA Animal and Plant Health Inspection Service, Plant Protection, and Quarantine (APHIS PPQ).

Under CSKT's Environmental Protection Agency (EPA) Program, the tribal, county, and federal governments agreed to assess where grasshoppers could be safely sprayed on the Flathead Reservation. CSKT GIS worked with these programs to analyze possible areas to spray based on tribal GIS data. Mapped results were then relayed to the spray pilot for applications. Funding for this project came from a Congressional appropriation through USDA-APHIS to manage lands, including land containing grasshoppers. A GIS solution was designed using field data collection, the buffer function in ArcGIS, interpretation, and maps.

The primary workflow for the Grasshopper Spray Project involved collecting locations of threatened and endangered habitat areas, such as bald eagle nests, using GPS in the field and other means. Data is stored in workgroup geogatabases on servers and generally organized by program. Once data is collected, the locations are buffered using ArcMap, and these areas become the GIS data of places to avoid in the spraying process. Interagency staff review and determine the quality control (QC) of the map, which results in the map to be used during the spraying process. The map review process also generates ideas on how to improve data and future data collection. Photo interpretation is also used during the QC stages with interagency staff and the CSKT GIS Program. National Agriculture Imagery Program (NAIP) imagery is used as a base layer in ArcGIS to identify additional boundaries of spray areas. Privately contracted spray pilots as well as on-ground spray personnel display the maps created from the CSKT GIS Program on their laptops or PDAs for guidance during the application of pesticides.

This project is ongoing. The parties involved have come back to re-implement the process for the past three years. The main challenge is accounting for all the sensitive areas in which not to spray. Another challenge is nonparticipating neighbors. If these neighbors do not spray or control grasshoppers by other means, a re-infestation of the grasshoppers occurs. To mitigate these challenges, CSKT solicited feedback and input from a variety of sources within the community.

The main benefit of this project is a significantly suppressed grasshopper population on the CSKT Reservation and in the surrounding community in Sanders County. Fewer grasshoppers means more forage and crops. Tribal agriculture businesses as well as nontribal agriculture businesses produce better yields than if the Grasshopper Spray Project had not been in effect. Increased forage for livestock results in less feed to purchase by tribal and nontribal ranchers. The interagency cooperation between CSKT, APHIS, and the Montana State University Extension was an extra benefit that improved the Grasshopper Spray Project beyond the scope of what CSKT could do on its own.

Photo by Kirsten Grish

9

Ensuring tribal safety

As sovereign Nations within the United States, tribal governments are responsible for the safety of all tribal members and communities living within the Tribe's jurisdiction. Many tribal public safety offices are using geospatial technology to facilitate safer communities within their jurisdictional areas. GIS is an essential technology for all disciplines that fall under public safety, whether it is emergency/disaster management, search and rescue, law enforcement, 911 dispatch, wildland fire management, or other applications. Many tribal governments face significant challenges ensuring public safety due to their limited resources and often distributed population. While GIS cannot make up for the lack of resources, it can significantly enhance the effectiveness of those involved with public safety both when responding to the Tribe's immediate needs and working to prevent future threats to the community. Each discipline within the public safety industry can benefit from a place-based solution, from preparation and mitigation to response and recovery. Rapid access to information, increased safety and efficiency, and better resource management are realized with the use of GIS. As the stories in this chapter demonstrate, putting spatial intelligence at the fingertips of the public safety staff and field personnel improves decision making and prioritization of efforts. Ysleta del Sur Pueblo and the Chickasaw Nation discuss how the use of geospatial technology can be a crucial tool when working to provide safe communities and well-organized responses to disasters and emergencies.

Using GIS to manage and recover from an unexpected wildland fire

Beau Barela, GIS Specialist, Ysleta del Sur Pueblo
El Paso, Texas

The present-day Tigua Tribe originated in the Pueblo of Isleta, just south of present-day Albuquerque, New Mexico. After the Pueblo Revolt of 1682, a small group of the Isleta Tribe was taken as forced labor and displaced to what is now El Paso, Texas. These tribal ancestors established Ysleta del Sur Pueblo. Since 1682, the Tigua Tribe has maintained a significant presence in El Paso and, despite much adversity, has persevered as a strong American Indian Nation.

Because Ysleta del Sur Pueblo is located in southeast El Paso County within the city of El Paso and the city of Socorro, a major issue for tribal leadership is the persistent encroachment by the surrounding cities. Part of the Ysleta del Sur Pueblo lands are adjacent to the US/Mexico border. Ciudad Juarez, Mexico, a city of over 1.3-million residents, is directly across the border from Ysleta del Sur Pueblo. The border location and surrounding metropolitan areas pose unique environmental and land management challenges to Ysleta del Sur Pueblo. The Ysleta del Sur Pueblo Reservation and properties are situated in a checkerboard fashion, with two reservation areas located approximately five miles apart. The Pueblo's land configuration is not contiguous and contains both urban and rural areas. Additionally, a 68,000-acre ranch was purchased outright by the Tribe and held through a simple fee title. Chilicote Ranch is located 200 miles southeast of the reservation in Jeff Davis County and Presidio County, Texas.

GIS efforts within the Environmental Management Office of Ysleta del Sur Pueblo began in 2005. GIS was primarily used for the inventory of tribal lands and for the air monitoring program. The use of GIS began at the same time the Environmental Management Office was established. According to the Tribe, GIS and environmental management naturally went together. The Tribe was awarded a two-year capacity building grant from the Administration for Native Americans (ANA) during the 2009 fiscal year. With this grant, Ysleta del Sur Pueblo was able to hire a GIS specialist and a GIS assistant. Since then, the impact of GIS at the Ysleta del Sur Pueblo has been significant. Some examples of how GIS is being used across the Pueblo include development of zoning maps for the Economic Development Department, crime mapping for the Tribal Police Department, resident parcel maps for the Housing Department, utility mapping for the Management Information System Department, invasive species mapping for the Environmental Management Office (EMO), wildlife inventory for the Wildlife Committee, pasture mapping for the Chilicote Ranch Management, member mapping for the Tribal Records Department, youth outreach with the Empowerment/Education Department, and Bureau of Indian Affairs (BIA) transportation strip maps for Tribal Operations and the engineering firm.

Ysleta del Sur Pueblo's first use of GIS for emergency management was in support of a wildfire at the Tribe's Chilicote Ranch. During the summer of 2009, a series of wildfires broke out at Chilicote Ranch. These fires threatened several ranch structures and infrastructure essential to the Tribes' cattle operations. The wildfire at Chilicote Ranch caught the entire Pueblo off guard. Since the Tribe purchased the ranch in 1998, no such event had ever happened. Once the EMO caught wind of the incident, several

Figure 1. The wildfire at Chilicote Ranch caught the entire Pueblo off guard. From Shutterstock, courtesy of Todd S. Holder.

of its members mobilized to help support in whatever way they could. Armed with a laptop computer, a GPS device (ArcPad), and a GPS camera, the EMO members left for the three-hour drive to Chilicote Ranch. Safety of the staff was the number one concern on everyone's mind, so the EMO was not able to immediately start collecting GPS points around the perimeter of the fire. It wasn't until early morning the next day that the conditions were right to start the collection process. While driving around the fire, it became clear that the dirt roads provided a natural fire barrier, along with the fire lines that the National Forest Service created. The actual collection process began by driving a little, taking a point, and continuing the process until the EMO was satisfied that it had mapped a majority of the fire. The EMO was able to create an accurate assessment of the fire damage by analyzing the collected GPS points, along with the natural fire barriers. Maps were created to provide fire responders, including tribal and National Forest Service responders with an accurate scope of the fire progression and damage. Once the fire was completely contained, the EMO was able to drive around the perimeter of the fire, gathering GPS points, and calculate the total damaged area.

The emergency response to the ranch fire was new to everyone from the Tribe; no design process existed for such an event. The collection of GPS fire points during the fire was, to say the least, chaotic. Once the EMO made it back to the office, the data analysis began. Information was needed on the total number of acres burned, the length of compromised fence line, and the length of compromised pipeline. There was not enough manpower or time to map the exact length of the fence and pipeline that had been damaged, so the EMO used a combination of new and existing data to estimate the damage.

The EMO knew the exact perimeter of the fire by importing the GPS points into ArcGIS and "connecting the dots" to create a feature class in ArcSDE. An attribute field was created to calculate the total acres. The wildfires had burned 6,610 acres of grazing pastures. Data for the fence and pipelines already existed in the spatial database engine (SDE) database, so calculating the compromised areas was a simple analysis performed by clipping features. The maps created by the EMO were used to aid in the insurance claim. Several miles of fence and pipeline were destroyed as a result of the fire, and the insurance company needed an accurate assessment of the damaged areas. Since the wildfire, GIS has been instrumental in the mapping of evacuation routes for the residents of Ysleta del Sur Pueblo.

This experience served as an eye opener for the entire tribal government. The EMO invested in more GPS devices and antennas. These antennas are magnetized and can be attached to any vehicle. In the event of another fire, the EMO can quickly mobilize and drive around the perimeter of the fire while at the same time keeping a track log of the driven area. The ability to stay inside a vehicle helps ensure the safety of all responders and personnel. Plans are in the works for an analysis of fire potential on the ranch using ArcGIS and image analysis/remote sensing software.

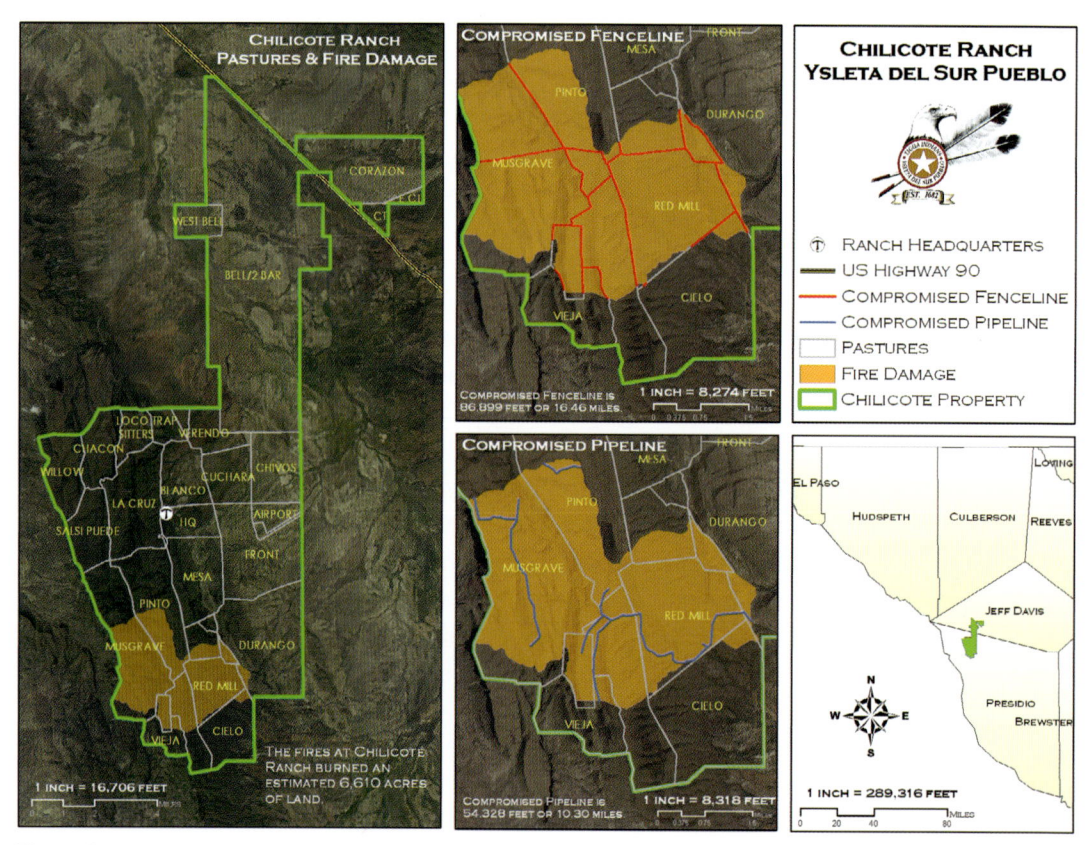

Figure 2. This map of the Chilicote Ranch wildfire was created for the insurance company, showing the total acres burned, the compromised fence, and the compromised pipeline. Courtesy of Ysleta del Sur Pueblo

Ysleta del Sur Pueblo uses an ArcGIS Server database to store all geospatial data. The ArcSDE product came as part of the ArcGIS Server package. The use of ArcSDE technology allows for a centrally stored set of data that can be disseminated across the entire organization. Along with ArcGIS Server and ArcSDE, ArcGIS Desktop, ArcPad, and ArcReader are also used.

The major challenges that were encountered for this project were lack of data, lack of manpower, and no existing protocols for a situation such as a wildfire. The ArcPad software and data needed to be loaded onto the GPS, which added to the dispatch time. The other issue was the scale of the topographic maps. A larger scale map was needed to include the full extent of the wildfire.

To overcome these problems, the EMO and GIS staff had to be resourceful with what they had on hand, using available technology by arming staffers who were available with ArcPad and GPS equipment, which allowed for an accurate assessment of the fire. There are now plans within the GIS division for such emergencies. The data is in place, ready for fast deployment into a mobile device armed with ArcPad. There is a special folder for the ranch; within this folder, there are topographic maps of all scales, vegetation maps, and maps showing the location of all structures. A 3D analysis of the ranch has been performed that is always available on laptops. This terrain analysis can be viewed in ArcScene.

As a result of this wildfire and project, Ysleta del Sur now has the ability to quickly provide valuable data to the Tribe's emergency responders. This information also allows for the quick dissemination of valuable data to all tribal leadership. The GIS staff is committed to ongoing analysis of the terrain and staying up to date on technology that may help in the event of an emergency. This project also helped the Tribe gain a better understanding of the terrain of the Tribe's ranch, and it is a benefit to know that in the event of an emergency, the Tribe can provide valuable information quickly and efficiently. The EMO would like the use of GIS to be a major part of all the decision-making processes by the Tribe, especially in emergency management, as GIS helps people make better decisions.

In addition to wildfire support, the tribal police department has used GIS to map crime hot spots on the reservation. Because of the Tribe's proximity to Mexico, special care must be taken in locating areas where illegal activity might be occurring. The EMO and tribal police use GPS to map areas on the reservation where the US/Mexico border fence has not yet been completed. These areas have been subjected to an increase in patrol.

Beau Barela, GIS Specialist for the Ysleta del Sur Pueblo, sums the project up by saying, "You may not always be prepared for every task that comes at you. GIS divisions within Tribes usually have to balance more than one task at a time. We don't have the luxury of only being charged with one task; we must provide data for the whole organization. When a problem comes at you that you are not completely prepared for, it's normal. Use the resources that you do have at your disposal and make the best of the situation at hand. Use other Tribes for technical support, ask your local governments for help, or contact Esri."

GIS and emergency management services at the Chickasaw Nation

John Ellis, Director, Department of Geospatial Information, the Chickasaw Nation
Ada, Oklahoma

The Chickasaw Nation has more than 49,000 members, and the boundaries of the Chickasaw Nation encompass more than 7,443 square miles in South Central Oklahoma, covering all or parts of thirteen counties. Because the Chickasaw Nation's boundaries cross several county lines, jurisdictional issues complicate challenges inherent in providing law enforcement and other emergency management services.

The Chickasaw Nation's Department of Geospatial Information (GSI) is committed to assisting and enhancing every element of emergency management services that operate within the Nation. Detailed maps of the service area in South Central Oklahoma are vital tools for the Chickasaw Nation Police Department (called the Lighthorse Police Department or LPD), the Chickasaw Nation Search and Rescue Team, and Bureau of Indian Affairs (BIA) firefighters. All offices that fall under Chickasaw Nation Emergency Management Services (EMS), as well as the Federal Emergency Management Agency (FEMA) and local county fire and police departments, benefit from maps and data provided by the GSI department. Hard-copy maps and map books of cross-deputized police and sheriff jurisdictions, zoning districts, and floodplains are a few of the key products the GSI produces for these offices.

Data storage is of great concern when dealing with EMS. Website and smartphone applications must be accessible to emergency responders at all times. All of the essential vector data that remains relatively static (roads, tribal lands, allotted lands, counties, the Chickasaw Nation Service Area, and so forth) is stored in a geodatabase that is structured in groups such as transportation, tribal holdings, administrative, and so on. The GSI staff also created a feature dataset specific for EMS that includes data layers for hospitals, fire stations, and police stations. All raster data is stored on two image servers named to describe what is within each folder, such as the Chickasaw Nation Service Area topographic maps, the 2008 National Agriculture Imagery Program, the 2009 4-inch imagery from Pictometry International Corporation, and so on.

Every separate project has its own folder on a network drive with a unique name under this system. The projects are named CNGIS_xxxx (for example, CNGIS_1543). Folders on the network drive are broken down to the division level, then the department, then a folder number. If a project is being worked on for the Lighthorse Police Department, the folder would be organized as

R:\GIS_Workspace\SelfGovernance\LighthorsePolice\CNGIS_1543.

Everything specific to a project will be saved within the project's folder. If a shapefile of a crime scene is needed for the project, it will be saved there. The ArcGIS project file will also be saved within that folder, along with detailed paperwork and anything else done for the project. Everything saved will have the same name as the folder (except shapefiles and similar items). The folder's contents may look like this:

CNGIS_1543.jpg, CNGIS_1543.mxd, CNGIS_1543.doc, CrimeScene.shp.

The GIS technician collects data for emergency management from multiple sources based on the project at hand. If the project yields data that is deemed to be of one-time use, it is stored in the correct

GIS folder on the network drive. Otherwise, it is passed on to the system administrator to be incorporated into the geodatabase following the standard operating procedure and potentially served out via an intranet website for emergency management services.

Points for all of the Chickasaw Nation facilities have been collected using GPS and placed in a geodatabase by GSI department staff. Photos, floor plans, and comprehensive attribute information have also been collected for each of these facilities. Access to this data provides additional resources for emergency management efforts.

This will be an ongoing project. Data is continuously updated and new information is entered. When a new map needs to be created for emergency management purposes, it is completed as fast and as accurately as possible. Unlike other areas of mapping, emergency management maps are usually needed as quickly as possible. They may need to be used in an upcoming court case, or they may be for a search and rescue effort, where time could mean the difference between life and death.

Numerous maps have been created and data has been collected, created, and stored for future projects. A website has been developed by the web applications developer and the GIS technician for the Lighthorse Police Department. This mobile web application uses ArcGIS Server. Since completion, the police officers are able to access the website from their patrol cars and are able to obtain crucial information on the spot. Officers are able to see if they are on tribal trust land. They are also able to locate the closest police stations, fire stations, hospitals, and other facilities in relation to their current location. Officers are able to access instantaneous information on Chickasaw Nation facilities, including photos, floor plans, number of employees, and critical imagery—all en route to the site of the incident. This is a tool that is used by Lighthorse police officers on a daily basis.

The GSI department continuously updates data and adds new information when available. GSI also incorporates new tools, as they are necessary or available, to ensure the safety of the Chickasaw Nation Service Area. A summer youth worker has been employed to provide support for EMS mapping. This position was used to create a Latta Schools map to plan for potential emergency situations.

Figure 3. The Lighthorse Police Department uses a mobile web application that uses ArcGIS Server to obtain critical information as they need it. From Shutterstock, courtesy of Phase4Photography

The continual benefit of this program is an efficient GIS that provides high-quality maps and data to help emergency services personnel do their jobs more efficiently and effectively. This helps make the Chickasaw Nation a safer place to live.

The project brought a greater realization of and appreciation for GIS and its diversity. All areas of emergency management operate more efficiently and effectively with GIS support. The project started with a few people from emergency management using GSI services. Once others saw the benefits of GIS, more departments and divisions within the Tribe began to use the department. Police and fire departments within the community that cannot fund a GIS are also using GSI services. The department workload continues to increase and exceed expectations.

In a perfect world, all local emergency management would operate with the GSI department's Esri-based GIS. Lighthorse police use the GSI department's GIS along with New World Systems' computer aided dispatch (CAD) system; the GSI department provides Lighthorse police with Esri shapefiles and Pictometry imagery to input into the CAD system. If everything could be done in the same coordinate system and projection, based on a unified schema, operations would be more seamless.

The Chickasaw Nation's GSI department offers this valuable advice: Tribes should realize all of the positive outcomes that will result from their GIS data. Natural disasters and emergency situations are unfortunate, but knowing that GIS will help save lives offers peace of mind. It is important to provide accurate data to emergency management workers for this reason. Providing quality data may take longer to create, but it will be worthwhile. Also, know that the job of an emergency management GIS worker is never complete. Know what data is available and what needs to be created. Don't be afraid to ask for help. There are many GIS companies that work with emergency management who have probably encountered some of the same problems and may have a simple solution that can save time and money. Also, be sure the emergency management personnel know that a GIS exists. Show them examples of what is possible with GIS. Let the emergency management personnel present examples of what would help them. Do a few projects for them so they will begin to see the benefits of GIS and feel comfortable using it.

to by Kirsten Grish

Supporting the tribal enterprise

Tribal governments are responsible for the stewardship of their lands and resources and for the health and well-being of their citizens. Tribal communities face diverse internal and external challenges and pressures in a complex and constantly evolving political and physical landscape. In order to make effective decisions, tribal leaders need to incorporate information from a multitude of sources to have sufficient context for the situations with which they are faced. GIS professionals are regularly asked to provide maps and data products to tribal leadership when the Tribe is faced with a complex situation. This should come as no surprise, as one of the inherent strengths of GIS is its ability to integrate information on a wide array of topics onto a common backdrop. Maps, however, are only as useful as they are accurate and up to date. Without an ongoing process, a system, to keep the information current and incorporate observations from a number of stakeholders, it is likely that basing a decision on an inaccurate map would do more harm than good. It is important to realize that by its very definition, GIS is intended to be used as a constantly updated and evolving *system*, not as a tool to conduct a linear process that leads to a static product. Systems are continuous, like the hydrologic cycle, and therefore, a GIS should be continually updated and maintained to support daily processes within the tribal government. In this way, when tribal leadership needs GIS output in support of key decisions, those products are as accurate as possible.

To achieve this decision-support platform, information from multiple tribal government offices must be considered. The flow of information from one tribal government office to another is inherently required for many tribal business processes. For example, a tribal member requesting a permit to drill a well or build a home may pass from the environmental, planning, historical preservation, natural resource offices and water code offices before being approved or denied. Site visits and the collection of related field data are typically required by multiple tribal government offices at various stages of the permitting review process. As this process unfolds, observations and information that are geographically tied to the location of the well are altered, which in turn affects the review underway in other offices. Other spatial data layers potentially maintained by each tribal government office will be compared against the well location during the deliberation on the permit request. GIS can play an essential role in streamlining business processes along these lines by providing an integrated platform, allowing each office to both manage certain spatial data for its own purposes, as well as contribute to a complete integrated platform for the tribal government. GIS platforms that integrate diverse stakeholder communities for a common operational purpose through defined workflows are often referred to as "enterprise GIS."

This flow of information from one office to another is required in tribal governments, yet many Tribes have yet to implement business systems to streamline access to key information in support of such processes. Today, a majority of Tribes use GIS in a given department, with limited staff for a limited purpose despite there being broad demand for GIS products and services across the government. GIS is a proven enterprise platform around the world for governments, companies, and institutions large and small that have recognized the importance of geographic information in their decision-making and business procedures. Increasingly, tribal governments are migrating toward enterprise use of GIS technology and in doing so are realizing a number of benefits, which include better access and flow of information across the government, broad consumption and stewardship of key information, tighter integration with other core business systems, increased operating efficiencies, and timely decision support.

GIS programs tend to follow a natural progression toward the enterprise. Once an organization comes to depend on geographic information products for decision support on a regular basis, the next logical step is to more tightly couple GIS with the broader tribal information technology (IT) framework so that information can be fused together in a timelier manner based on defined rules and processes. In a tribal government, demand for maps is typically high across all government departments or business units. This forces many GIS staffers to be constantly reacting to requests for maps, rather than conducting meaningful analysis or developing or maintaining a shared GIS infrastructure. The solution to this challenge is enterprise GIS, an integrated platform that allows multiple tribal offices to be the data stewards of their own data as defined by their core mission within the government. Enterprise GIS enables the entire tribal government to access GIS products and services on demand while providing a consistent platform for updating geographic information in support of regular business processes.

Each Tribe has unique requirements and organizational structures. The first step in designing any enterprise GIS is understanding these diverse requirements. While there are many digital and print publications available that define the process of gathering user requirements, common elements include interviews with key stakeholders and the evaluation of the information products they need to do their work. From these requirements, a GIS software architecture can be designed to meet these requirements. One of the core aspects of any enterprise GIS solution architecture is data management, which involves how each office will access the data needed to conduct their work. This is where the use of geographically enabled databases (or geodatabases) becomes invaluable. While there are individual geodatabase solutions, workgroup and enterprise geodatabases are designed to support potentially large numbers

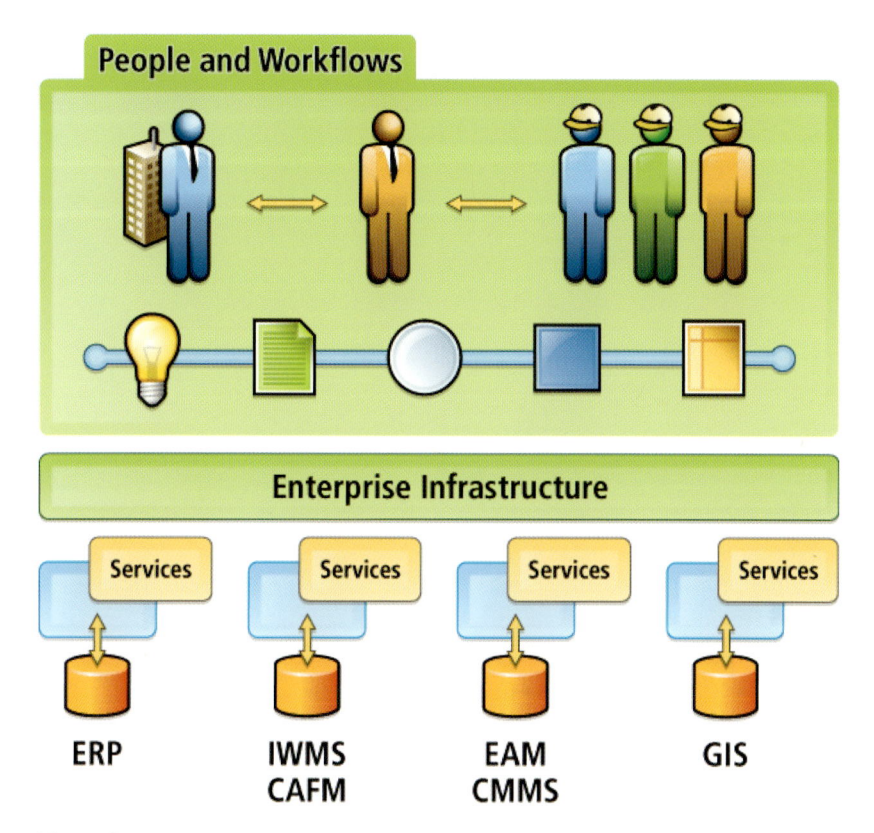

Figure 1. Enterprise GIS conceptual framework.

of data stewards, each with unique read and write privileges. As the case studies in this chapter show, implementing a geodatabase alone brings terrific improvements to information management. In many cases, tribal government offices will simply need to view the GIS data to support their activities. In other cases where the office needs to update or change information in the geodatabase, these groups can become the stewards of that information and can be provided the appropriate permissions to manage their own data. Through a security policy, typically managed through the IT department, the ability to view and update data in the enterprise system is established and enforced. Defining the steps involved in updating the information in the enterprise GIS can be documented as workflows, which can be reflected in the geodatabase design to assure required information is collected during the editing process. Documented workflows are valuable within tribal government offices because they define a standard operating procedure, which greatly streamlines updating the system as well as reduces the training investment required for bringing on new staff. For example, a workflow for permitting can define the flow of the permit application through the Tribe. This workflow can be implemented within the GIS so that the individual tasks in the process are tracked and participants are notified, or reminded, when they are required to comment on the permit to allow it to move forward in the process. Implementing workflows along these lines are streamlined by the use of the ArcGIS Workflow Manager extension that provides this capability.

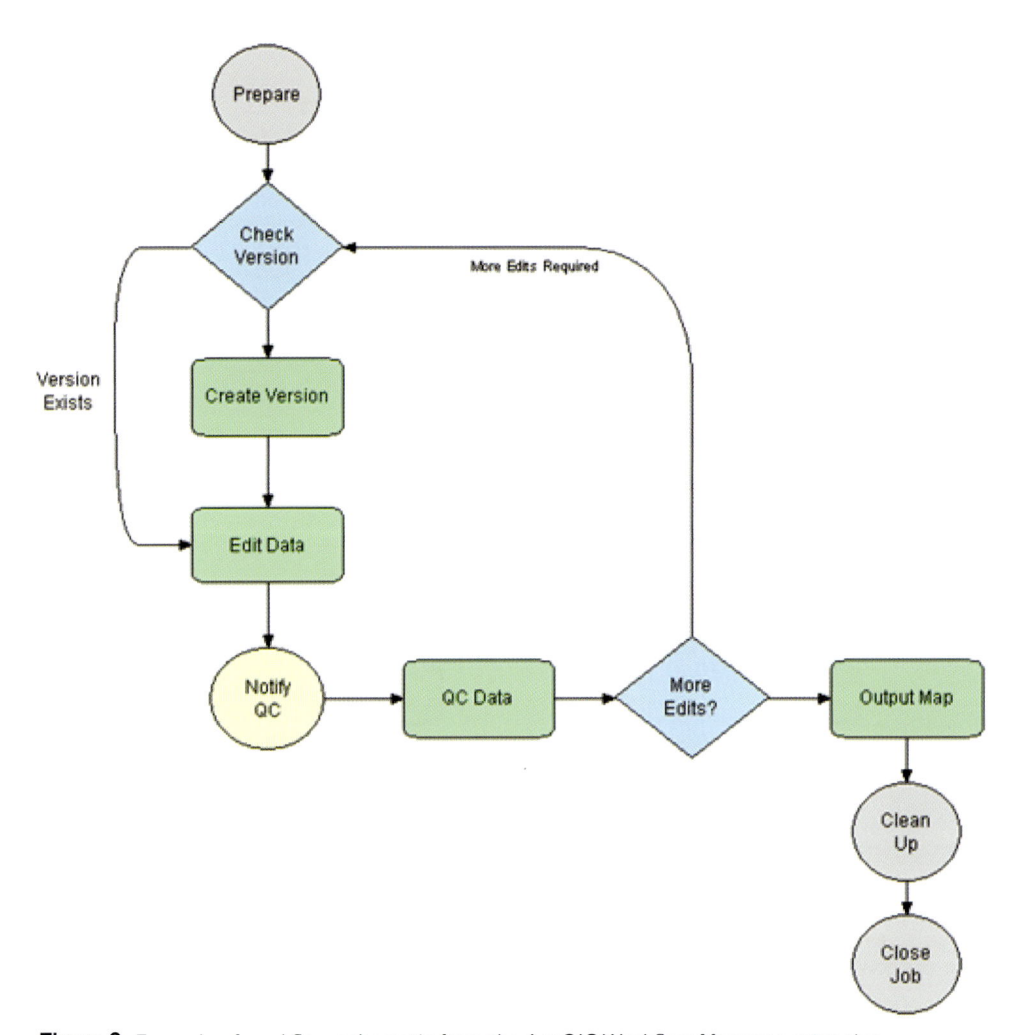

Figure 2. Example of workflow schematic from the ArcGIS Workflow Manager extension.

Another fundamental concept of enterprise GIS systems is the access to live data through web services. Rather than print a map or export a file from the geodatabase, a web service is a live view of the data that can be accessed through a number of clients, including a web browser. Using web services, the tribal staff is not required to learn complex GIS software but is able to access maps and update files through a simple internal web application that permits viewing the current data from the geodatabase. Enabling geographic information through web services is a core capability of the ArcGIS platform. Publishing geographic information through web services allows this information to be viewed and updated on a wide array of devices, including mobile phones and handheld GPS devices. In this way field staff, law enforcement, and tribal leaders can have access to key information at their fingertips. These primary elements of an enterprise GIS, when coupled together, provide an integrated information management platform for the Tribe.

Implementing enterprise GIS, however, is not a trivial task; it requires thoughtful system planning and investment. As other national governments around the world have faced similar information management challenges across diverse stakeholder communities, the concept of spatial data infrastructure has emerged as a way to define the business logic of enterprise GIS. GIS software is just one component of a spatial data infrastructure. Timely decision support and improved citizen services require accurate data, standards, business processes, institutional arrangements, information systems, and enabling policies and legislation to facilitate the flow of critical information through the Tribe. The concepts of enterprise GIS and spatial data infrastructure reflect this collection of technology, standards, and policy to enable the flow of geographic information across a government to eliminate information silos. To achieve an enterprise GIS solution, investments may be required in system design, regular training and capacity building, hardware, software, and potentially, staffing. Still, in the return on investment analysis across the GIS industry, time and again this investment of GIS at the enterprise level provides better transparency and accountability of daily business processes, leading to better efficiencies and cost savings.

This chapter showcases five Native American communities that have made the investment in enterprise GIS as a platform for more efficient tribal government. They are the Ysleta del Sur Pueblo (Texas), the Agua Caliente Band of Cahuilla Indians (California), the Eastern Band of the Cherokee Indians (North Carolina), the Confederated Tribes of Grand Ronde (Oregon), and the Chickasaw Nation (Oklahoma). The stories in this chapter illustrate that while each community has taken a unique path to GIS at the enterprise level, achieving enterprise GIS does not require a large program, large volumes of staff, nor decades of experience with the technology. In the first example, the Ysleta del Sur Pueblo makes significant advances toward the enterprise in only one year beginning in 2009.

Enterprise GIS fast track
Beau Barela, GIS Specialist, Ysleta del Sur Pueblo
El Paso, Texas

The Tigua Indians originated in the Pueblo of Isleta, just south of present-day Albuquerque, New Mexico. After the Pueblo Revolt of 1682, a small group of the Isleta Tribe was taken as forced labor and displaced to what is now El Paso, Texas. These tribal ancestors established Ysleta del Sur Pueblo. Since 1682, the Tribe has maintained a significant presence in El Paso and despite much adversity has persevered as a strong Native American Nation. Ysleta del Sur Pueblo is located in southeast El Paso County within the city of El Paso and the city of Socorro. A major issue for tribal leadership is the persistent encroachment by the surrounding cities. Part of Ysleta del Sur Pueblo is adjacent to the US/Mexico border. Ciudad Juarez, with over 1.3-million residents, is directly across the border from Ysleta del Sur Pueblo. The border location and surrounding metropolitan areas pose unique environmental and land management challenges to Ysleta del Sur Pueblo. The Ysleta del Sur Pueblo Reservation, which includes both urban and rural properties, is situated in a checkerboard fashion, with two reservation areas located approximately five miles apart. Additionally, a 68,000-acre ranch was purchased outright by the Tribe and held

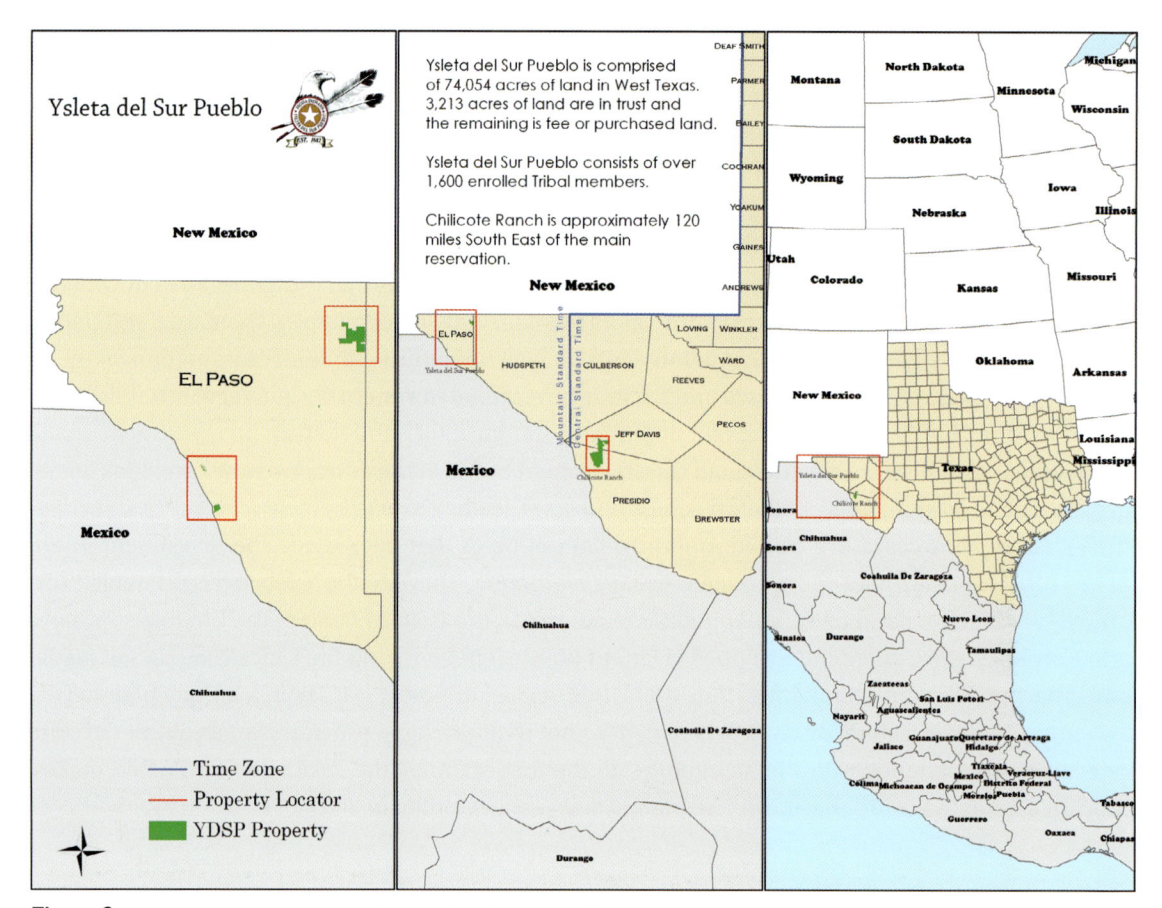

Figure 3. Overview of current tribal lands of the Ysleta de Sur Pueblo. Courtesy of Ysleta del Sur Pueblo

through a simple fee title. Chilicote Ranch is located 200 miles southeast of the reservation in Jeff Davis and Presidio Counties, Texas.

In 2009, Ysleta del Sur Pueblo was awarded a two-year capacity building grant through the Administration for Native Americans. Before the Tribe was awarded the grant, the Environmental Management Office (EMO) was the only tribal entity directly using GIS. Several departments contracted engineering firms to create maps for various tasks, but no work was done internally by tribal staff. The EMO used GIS and GPS to inventory tribal lands and to help with the Tribe's air monitoring program. When tribal government leaders first learned about the capabilities of GIS and GPS, they showed excitement about the program and its potential. Tribal leaders saw how the use of GIS and GPS would be useful tools in providing data for both the tribal government and the Ysleta del Sur Pueblo community.

The EMO hired a GIS specialist and a GIS assistant to take over all GIS duties for the Tribe. Beau Barela was hired as the GIS specialist, and Andrea Everett was hired as the GIS assistant. The GIS division first started storing data in file geodatabases. The popularity of GIS soon grew, resulting in a need to migrate from the file geodatabase to an enterprise-grade geodatabase. A Microsoft SQL Server database coupled with Esri ArcSDE technology was the logical solution to the data distribution problems.

The combination of demand for GIS data among the entire Pueblo and the availability of Esri software helped move Ysleta del Sur Pueblo from a file geodatabase system to a more advanced and efficient enterprise GIS platform.

As the use of GIS within the EMO became more widely known, other tribal government offices expressed interest in GIS products and services. Recognizing that the office did not have the resource or mandate to provide all of the GIS services to all of the interested offices, a system design process was undertaken to determine an enterprise architecture that would serve the Tribe's organizational needs. As a first step, a user needs assessment was completed to gather requirements from each office for the design process. This needs assessment involved interviewing each department director about the level of use they expected from the GIS, the type of data they needed access to, and the type of analysis and reporting they wanted to perform.

In time, a number of tribal government offices were supported by the enterprise GIS platform, including tribal police, environmental, economic development, housing, and emergency management. Each department is connected to the enterprise geodatabase, and each is responsible for its own GIS layers or feature dataset under the main geodatabase. This allows for easy organization of all GIS data within the geodatabase. Permissions are set based on the needs of each department. Read-only rights were given to all users for base layers, whereas the departments have full permissions to their own data. Feature datasets were created to separate different types of data into easily manageable areas. Permissions are provided to each department according to their needs. Risk was minimized during the transfer from file geodatabases to the enterprise geodatabase by ensuring that proper systems backups were in place. The Tribe's Management Information Systems (MIS) Department was critical in the initial setup and backup of this database.

One of the primary problems that this enterprise GIS platform helped solve was allowing for multiple users to edit the data simultaneously. The Tribe's housing department needed an effective way to update its housing information on a daily basis. In addition to needing to keep its own data up to date, environmental GIS layers managed by the EMO were needed by the housing department to support its business processes. Before the enterprise geodatabase was in place, the housing department would have to write its updates on a paper map and send the map to the GIS Division for updates. Now that the enterprise geodatabase is in place, the housing department has the ability to update its data as needed and view housing data along with the current environmental layers.

When the enterprise framework was put into place, an effort was undertaken to incorporate as much tabular data as possible into the enterprise. In most governments and businesses, important business information is stored in tables of information such as Microsoft Excel. This is effective for small static tables, but this document type does not readily support the exchange or update of information across a larger stakeholder community. To remedy this, participating tribal government offices were interviewed to determine what tables of information they were currently managing could be referenced and managed in the enterprise GIS. Once the tabular data was referenced to its associated location in the GIS, or georeferenced, it was then loaded into the enterprise geodatabase. Appropriate permissions were then set up to dedicate the primary department to manage the GIS layers. Providing each department access to the enterprise system to modify and update the data as needed resulted in fast, efficient, and effective management of tribal information.

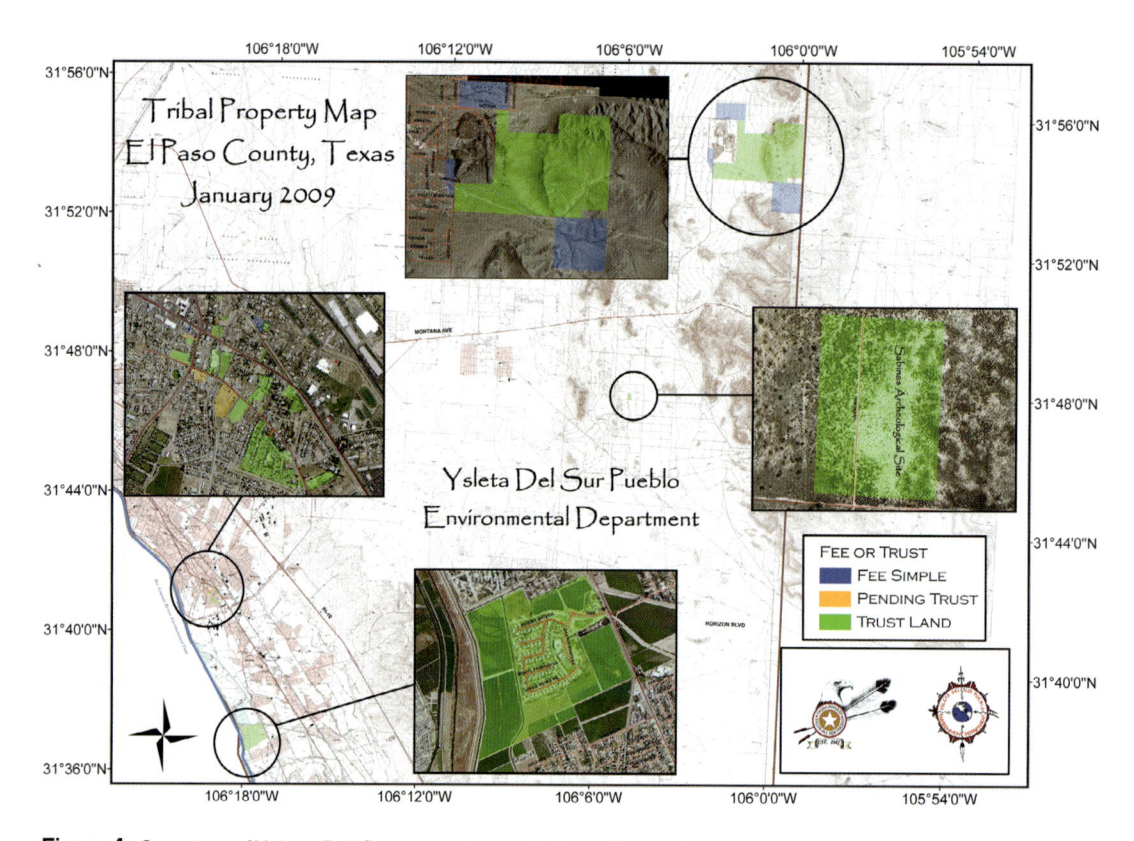

Figure 4. Overview of Ysleta Del Sur properties in El Paso County Texas. Courtesy of Ysleta del Sur Pueblo

One of the lessons learned in the implementation of enterprise GIS at the Ysleta del Sur Pueblo is the critical relationship between the GIS and the IT or MIS Departments. MIS has an important job in securing data and ensuring proper protocol is being followed when handling tribal data. As the GIS and MIS departments won't always see eye to eye, it is extremely important to have good communication between the two departments. Flexibility is very important while implementing enterprise GIS. Since establishing the core enterprise GIS architecture, the next step in the enterprise evolution is to leverage ArcGIS Server to share data and products further across the tribal government and with the tribal community.

With the installation of the GIS enterprise system, different departments have the opportunity to become trained in GIS technology to become effective data stewards of their information. With the enterprise GIS, they can modify and update their data as needed, without having to wait for the GIS office to apply the necessary changes. Enterprise GIS technologies have been instrumental in helping tribal leadership gain a better understanding of their lands and spatial relationships happening on those lands. The EMO sees the use of enterprise GIS as not only an improvement in the flow of information management at the Tribe but as a pathway for creating jobs and other opportunities for tribal members.

Enterprise geodatabase design

Volker Mell, GIS Coordinator, Confederated Tribes of Grand Ronde
Grand Ronde, Oregon

In chapter 1, the Grand Ronde Natural Resources Department (NRD) described how it had implemented GIS at the enterprise level to effectively manage its sustainable forestry programs. The Tribe's GIS program, however, reaches far beyond the NRD and effectively serves many other offices within the tribal government. This was achieved through the thoughtful design of an enterprise-grade GIS platform arranged around the specific needs of each office. In order to appreciate how this was achieved, consider a brief history of the Tribe's GIS program. In this case as well as in that of Ysleta del Sur, achieving enterprise GIS does not require a decade of GIS work, nor does it require large numbers of staff.

In 2007, after two years without a GIS analyst or any GIS expertise at the Confederated Tribes of Grand Ronde (CTGR), the Tribe decided to do a detailed analysis on its need for GIS. A contractor who conducted the needs assessment recommended the establishment of an enterprise GIS under the direction of a GIS coordinator. Through general funds, an enterprise GIS program was established for the tribal government. The GIS program consists of one GIS coordinator who was hired in July of 2007.

One of the initial tasks undertaken by the GIS coordinator was a more detailed user needs analysis following the concepts outlined in *Thinking About GIS* by Roger Tomlinson (2005), considered by many to be the "grandfather of GIS." The result was then presented to the departments at CTGR. The needs analysis showed that the primary GIS stakeholders were the natural resources, cultural resources, planning, engineering, and housing departments. At the time of the survey, about twelve staff members used GIS on a regular basis, with most using ArcGIS Desktop. The survey identified that the most intensive user base was the NRD. The survey also revealed that despite earlier efforts by the former GIS analyst (in 2005) to establish a centralized geodatabase, GIS data was stored in several data silos across the Tribe. The existing layers had four different projections, and 40 percent of the GIS layers did not have any projection information.

As a result of the user needs analysis, the following architecture was established by the GIS coordinator. ArcGIS Server was used to establish four central geodatabases:

- A geodatabase storing grid based data (DEMs, aerial photos, satellite data, and so forth), which does not need to be edited
- A database storing vector data of interest to all CTGR GIS stakeholders (roads, contour lines, city limits, highways, counties, and so forth), which only gets edited by the GIS coordinator
- A geodatabase storing culturally sensitive data (archeological sites, historical trails, and so on), which gets edited through versioned editing by the cultural resource department
- A geodatabase storing the GIS data layers pertaining to the NRD (cut blocks, forest stands, local hydrographic data, and so on), which is edited on a regular basis by the NRD staff through a hierarchical system of versioned editing in ArcSDE

Combined, these geodatabases hold about 100 GB of content and are backed up on a daily basis. All feature classes in the databases are mapped to layer files in a file server system, which predetermines symbology, and any necessary data joins to external tables. The users only need to load the layer files

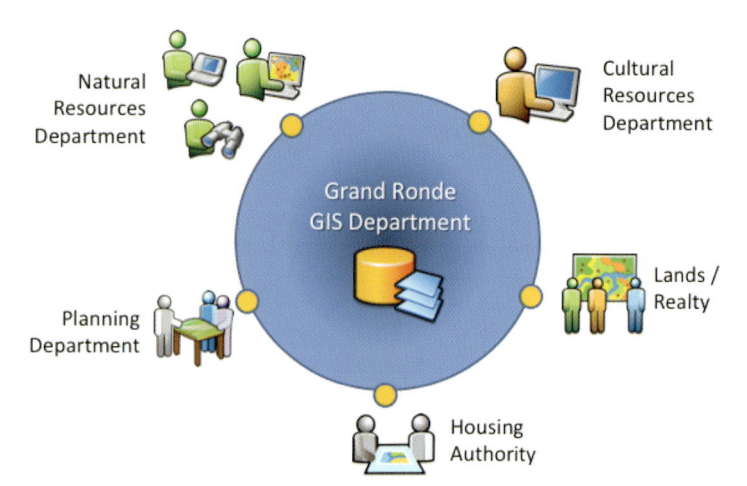

Figure 5. Conceptual overview of Grand Ronde enterprise GIS.

rather than having to deal with the layer naming conventions in the geodatabase. All users are connecting to the geodatabases through operating-system-based authentication. The CTGR-IT department organizes the users into several GIS user-groups after directions from the GIS department.

Training was one of the main efforts during the first year of the development process. This was essential in assuring the appropriate use of the enterprise architecture. The GIS coordinator provides regular training on the use of ArcGIS Desktop to the participating tribal government offices that are now capable of solving day-to-day GIS tasks without the help of the GIS department. Classes were offered on a regular basis (every two weeks) with different topics in an adult learning center computer room.

The NRD is using ArcPad and the checkout process to edit geodatabase layers in the field. The workflow from geodatabase to ArcPad field data and back has replaced a cumbersome process of finding geodata scattered in several locations on file servers and not knowing the projection or how current the data is.

In a later phase of the enterprise GIS development, the GIS department has been implementing ArcGIS Server to reach GIS stakeholders whose GIS needs could be addressed through web-based applications. Two web applications focusing on the status of tribal lands have been developed for the lands department. Besides benefiting the lands department, these applications are also accessible by other CTGR government employees. The tribal council and the tribal government executive office are benefiting from these applications by being able to identify tribal property and potential future land acquisitions. These web applications as well as most of the smaller scale ArcGIS Desktop maps stored on the system as template maps use ArcGIS Online basemap services for the background aerial photo imagery. A fast Internet connection makes it possible to use these services effectively. The GIS department views the planning and development of the enterprise GIS as a permanent effort to keep the GIS up to date and adjusted to changing user needs and expectations.

References

Tomlinson, Roger. 2005. *Thinking About GIS: Geographic Information System Planning for Managers*. Redlands, California: Esri Press.

Sustaining economic development on sacred lands
Beckie Howell, GIS Manager, Agua Caliente Band of Cahuilla Indians
Palm Springs, California

In chapter 4, the Agua Caliente Band of Cahuilla Indians was featured, describing its advanced realty program. The Tribe's enterprise GIS architecture provides spatial data services dynamically as live web services to six different tribal government divisions, including planning, economic development, realty, construction, historic preservation, and government affairs. More than ten tribal government offices consume products from the GIS platform in support of their daily activities. These include planning, economic development, realty, construction, historic preservation, canyons, government affairs, tribal programs, tribal administration, and tribal education services.

While each Tribe has unique challenges and requirements, there are some common elements of enterprise GIS systems. One core element is the geodatabase. While it is perfectly appropriate for individuals who work alone on GIS projects to store their GIS layers as files or in a personal geodatabase, as soon as the data being managed in the GIS has implications beyond that individual, an enterprise geodatabase is a more suitable data management platform. Enterprise geodatabases are designed for both storing and managing GIS data across diverse communities of editors and viewers while maintaining the integrity of the data. Because the geodatabase is in effect a geographically augmented relational database management system, all of the benefits for compression, backup, versioning, and performance are extended to the GIS system through this model.

In the case of the Agua Caliente, the migration to an enterprise geodatabase was driven by the fact that tribal government offices did not have access to the GIS data they needed to support their mission. Thus, the driving factor to migrate to an enterprise geodatabase was data access and scalability. The GIS group was having increasing problems with multiple users being locked out of GIS layers due to limitations of the personal geodatabase, which is not optimized for multiple viewers or editors, unlike an enterprise geodatabase. As a result, workflows were being disrupted. In addition, the sizes of the databases were growing to an extent that we were filling personal geodatabases to capacity. There was also a desire to expand the GIS audience within the organization, so issues of performance became a key consideration. Enterprise geodatabases are optimized to handle numerous users without affecting performance.

As was done by the other Tribes described in this chapter, a user needs exercise was conducted to determine the technical requirements of the system and which stakeholders would be served. This design process included a two-day long discussion with IT staff, the GIS manager, and a representative from Esri. The discussion included evaluation of the existing GIS data and databases, server configuration, migration expectations, and a needs assessment. The GIS manager also spoke with the other tribal division managers to determine how they could see GIS providing a service to their departments. The GIS manager then used this feedback to draft a requirements document that contributed to the design of the system.

The resulting enterprise GIS system at the Agua Caliente Band of Cahuilla Indians includes an enterprise geodatabase based on Microsoft SQL Server. Three physical servers powering two enterprise geodatabases are used, including one for the databases and SQL Server; one as the application server

154 | Tribal GIS: Supporting Native American Decision Making

with ArcGIS Server, Image Server, and the ArcGIS Server Object Container (SOC); and one as the web server with ArcGIS Server, Image Server Manager, and the ArcGIS Server Object Manager (SOM). In this architecture, ArcGIS Server components are distributed across multiple physical servers to meet the unique performance needs of the Tribe. Each geodatabase serves a unique purpose and stakeholder community. The production database is where the GIS group edits and creates data. Once data has been updated and verified, it is pushed to the consumption database for the GIS users to access. Within the

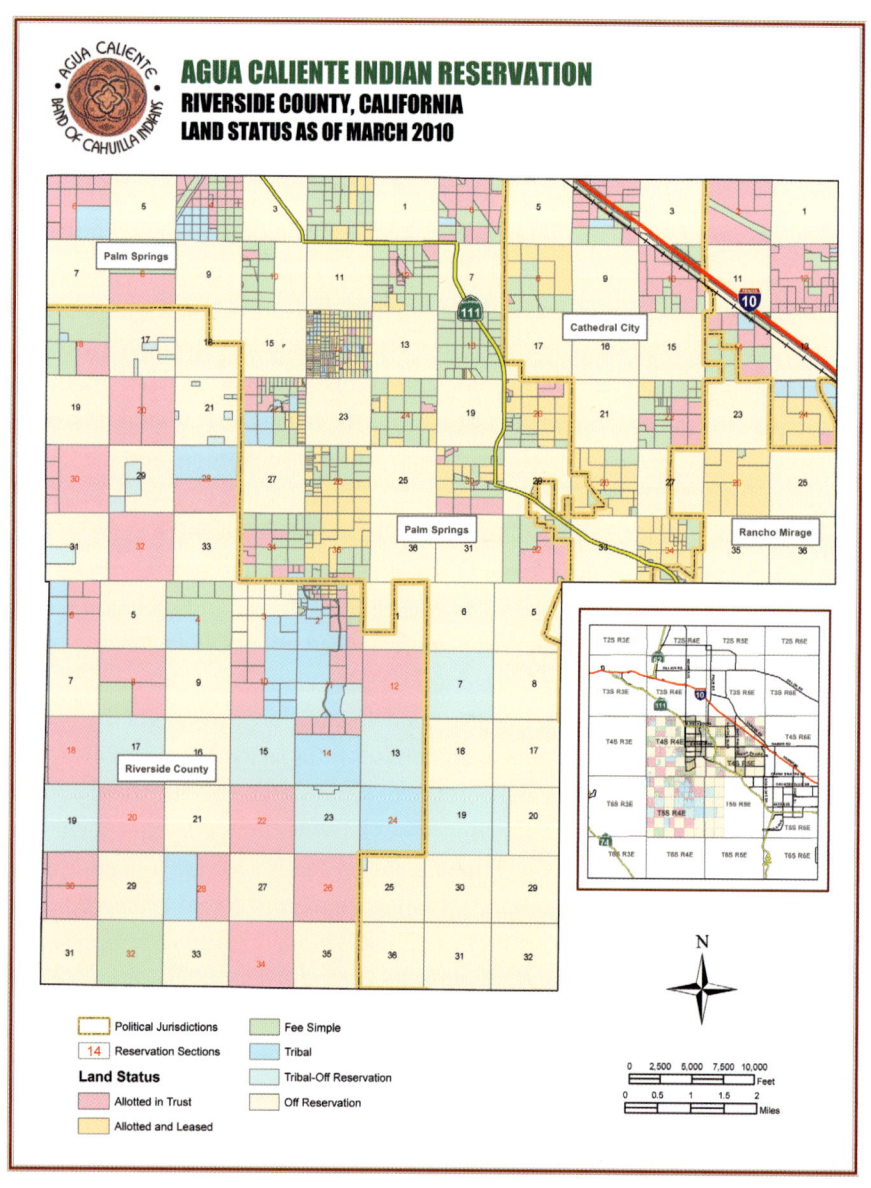

Figure 6. Status of tribal lands of the Agua Caliente Band of Cahuilla Indians.
Information and graphics provided by the Agua Caliente Band of Cahuilla Indians

production database, there are seven different database schemas created to organize the data by type and geographic projection. An archive schema stores old versions of data for historic purposes. In total, the enterprise geodatabases contain nearly thirty gigabytes of vector data and approximately one terabyte of raster data. As data is updated, it is pushed from the production database to the consumption database, so the GIS users always have the most current data.

It is important to note that the migration of the new system was approximately a week-long process. After the new enterprise geodatabase was established, the GIS group analysts continued to use the old databases while the GIS manager migrated the data and tested the functionality of new databases. The testing included viewing, editing, and moving data, as well as database administration tasks. Once it was determined that the new enterprise databases were fully functional, the GIS analysts began accessing and using the new geodatabase, and the migration was complete.

This advanced enterprise GIS platform is managed primarily by two staffers—the GIS manager and a GIS analyst. The GIS manager acts as the database administrator and takes the lead on major mapping projects. She is responsible for managing the department budget, designing and overseeing the databases, and assigning tasks to the GIS staff. The GIS analyst handles all day-to-day mapping requests of the tribal organization, as well as taking on long-term mapping and analytical projects. By embracing enterprise GIS technologies, these two individuals make a significant contribution to information management at the Agua Caliente Band of Cahuilla Indians.

An integrated geographic information system

David Wyatt, GIS Manager, Eastern Band of the Cherokee Indians
Cherokee, North Carolina

The Eastern Band of Cherokee Indians (EBCI) is the only federally recognized Tribe in North Carolina. It is located in the southern Appalachian Mountains. Historically, the Cherokee Tribe was one of the largest Tribes east of the Mississippi, covering approximately an eight-state region in the Southern United States. For the first 200 years of contact with Europeans, the Cherokees extended hospitality and help to the newcomers. Peaceful trade prevailed. Intermarriage was not uncommon. The Cherokees were quick to embrace useful aspects of the newcomers' culture, from peaches and watermelons to written language. This last was single-handedly created by the Cherokee genius Sequoyah, who introduced his "syllabary," or Cherokee alphabet, to the Cherokee National Council in 1821. Within months, a majority of the Cherokee Nation became literate.

But by then, nearly 200 years of broken treaties had reduced the Cherokee empire to a small territory, and President Andrew Jackson began to insist that all Southeastern Indians be moved west of the Mississippi. The federal government no longer needed the Cherokees as strategic allies against the French and British. Land speculators wanted Cherokee land to sell for cotton plantations and for the gold that was discovered in Georgia. Although the Cherokees resisted removal through their bilingual newspaper and through legal means, taking their case all the way the Supreme Court, Jackson's policy prevailed. In 1838, events culminated in the tragic Trail of Tears, the forced removal of the Cherokees in the East to

Oklahoma. One-quarter to one-half of the 16,000 Cherokees who began the long march died of exposure, disease, and the shock of separation from their home.

The Cherokees in Western North Carolina today descend from those who were able to hold on to land they owned, those who hid in the hills, defying removal, and others who returned, many on foot. Gradually and with great effort, they have created a vibrant society, a sovereign Nation of 100 square miles where people in touch with their past and alive to the present preserve timeless ways and wisdom.

A fundamental role of any tribal government is to provide optimized services to its constituents—including planning, development, project management, permitting, public works, public services, cultural preservation, environmental protection, legal records management, and land records management through time. As with many tribal governments, providing efficient and accurate services to tribal members has been a challenge for the EBCI. In a major effort to improve services to tribal members, the EBCI has invested in an integrated GIS (or IGIS). The IGIS promotes the principal chief's goal of sustainable development projects in the Cherokee community by establishing and implementing technology management information systems to assist with the effective and efficient administration of development processes in tribal government programs.

The EBCI saw the need to develop the IGIS for a number of reasons:

- Growing pains. First, the Tribe had experienced significant economic growth, and there was a need for better integration of government services. Second, the Tribe sought to improve organizational integration—tribal programs are currently limited, with no single location to store, track, update, share, and report on project status and integration with the tribal IGIS. One of the main benefits of GIS is improved management of organization and resources. A GIS can link datasets together by common location, such as addresses or spatial location, which helps departments and agencies share their data. By creating a centralized database, one department can benefit from the work of another. Data can be collected once and used many times—without the concern that the information is being duplicated.

- Better decisions. Short- and long-term benefits of combining existing EBCI business processes into a GIS-centric IGIS geodatabase include tracking, reporting, enhanced productivity, efficiencies, and cost savings to better manage GIS tribal resources. The adage, "better information leads to better decisions," is true for GIS. A GIS is not just an automated decision-making system but a tool to query, analyze, and map data in support of the decision-making process. The IGIS improves the site plan review process to ensure all internal and external EBCI processes are performed and addressed accordingly. Regarding the site plan review process, the information gathered early on in the process can be shared with multiple downstream departments to benefit their use of already acquired spatial and tabular/attribute membership data. This sharing of data will ensure nonduplicated and more correct membership information and will provide data analysis functionality not currently available within the Tribe. The IGIS will also uniquely benefit the tribal government when it takes over the land management and site survey processing currently being performed by the Bureau of Indian Affairs (BIA).

- Data silos. The Tribe also recognized the risk of data silos and duplication. Traditionally, government agencies begin GIS use on a departmental level in order to meet their own internal business needs. Since data is often not readily available in a GIS-compatible format, each department begins to

build its own silo data holdings. The focus is typically on specific data required to meet the primary business needs of the department, but it is quickly determined that ancillary data is also required to provide the necessary context for analysis. Since departments are working internally and their GIS requirements are constrained to their own particular business area, cross-departmental data sharing is not always possible during the early stages of GIS implementation. This frequently results in the creation of "data silos" and duplication of data.
- Making maps. Making maps with GIS is much more flexible than traditional manual or automated cartography or computer assisted design (CAD). A GIS creates maps from data located in databases, which are kept up to date through the enterprise system. Existing paper maps can be digitized and translated into the GIS as well. The GIS-based cartographic database can be both continuous and scale free. Map products can then be created centered on any location, at any scale, and showing selected information symbolized effectively to highlight specific characteristics.
- Value of GIS. The Tribe recognized the value of GIS as a planning and decision-making tool. Several departments throughout EBCI are currently utilizing GIS on a daily basis and have extensive data holdings. Other departments that are not currently making full use of GIS have identified a number of business areas where they believe implementation of GIS will have a positive impact on their processes and procedures.

The primary purpose of the IGIS is to serve as the central repository of geographic information-related tribal government services. This central repository includes a defined business workflow to update and maintain the Tribe's land record management system. The IGIS system supports a number of web services and web-based applications for real estate evaluation, land records management, economic assessment and planning, and building site project review, inspection, management, and permitting. Each of these applications integrates business requirements and evaluation by EBCI departments such as housing, utilities, engineering, and cultural and environmental resources. The IGIS also provides integration of business workflow to support archeological and environmental resource management, disaster planning, and EMS and e911 services.

The IGIS resolves the problems associated with data silos by providing and implementing an enterprise geodatabase that serves as a central repository for all GIS data. Each user throughout the enterprise has access to the most current, verified, and accurate data available. When data is required for a project, it can be accessed directly from the centrally maintained GIS, ensuring the highest level of confidence in the analysis results. Furthermore, all data is stored in a common schema that meets the business needs of the entire enterprise. A geodatabase design is developed based on enterprise-level requirements provided by the core GIS users in each department. This common schema ensures that users know exactly what is available in the GIS and that it will be in the same format each time it is accessed. A comprehensive data maintenance plan, complete with user roles and permissions, ensures that data is maintained and updated only by those individuals and agencies most appropriate to "own" the data. The utilities department, for example, would have ownership and maintenance responsibilities for water and sewer data while the Cherokee Department of Transportation (CDOT) would maintain street centerlines and rights-of-way.

The IGIS fulfills one of the fundamental governmental roles of the EBCI by providing efficient services to the Cherokee community. Specifically, the IGIS is designed to support EBCI by meeting the following requirements:
- Transparency of tribal government projects and services to the Cherokee community.
- Interoperability and integration with an electronic document management system (EDMS).
- Efficient data maintenance standards by requiring all contractors to deliver project data in specific compatible GIS data format that meets the IGIS schema.
- Integration with other primary tribal government systems, such as contract management, lease management, and services applications (housing, management, project management, federal/state permitting coordination).

Historical documents relating to the Tribe's boundaries and land administration are essential artifacts in managing tribal lands. EBCI was the first Tribe to obtain access from the US Department of Interior, BIA, to all of its land possessory hold documents. The BIA provided access to these records in an effort to work with EBCI to refine the antiquated paper systems used currently to administer land records. EBCI then digitally scanned over 1.2-million historical documents and loaded/indexed them into an EDMS. Through a contracted engagement with Esri Professional Services, these EDMS records are being used to reconstruct the 40,000 splits, mergers, and changes of EBCI tribal lands to develop a cadastral fabric for the Tribe's lands. A cadastral fabric is a topological database model that is designed to streamline transactions common in land records management activities. The EDMS records are linked to the IGIS parcel records and these records are in turn linked to the EBCI financial system of record. The goal of the IGIS is to be integrated and interoperable with all of EBCI's major departmental data repositories. The IGIS platform also includes a number of web-based applications to replace and streamline over a hundred paper form-based business workflows that are all ultimately tied to the IGIS land records repository.

Due to the unique nation-to-nation relationship with the federal government, Tribes are in constant dialog with most federal agencies. While this dialog dates back to the earliest origins of the United States, areas of improvement remain in the dialog between Tribes and most federal agencies. One benefit of the EBCI investment in the IGIS platform is to significantly streamline the exchange of information with state and federal agencies. Through leveraging the same system design and security principals to tightly define access and the read-write privileges of each stakeholder, the IGIS platform now effectively brokers information exchanged between the Tribe and a number of state and federal agencies, including the BIA, Bureau of Land Management, Environmental Protection Agency, Army Corps of Engineers, US Fish and Wildlife Service, and North Carolina Department of Environment and Natural Resources.

This structured government-to-government exchange of information through the IGIS has improved the management of water quality, wetlands survey, flood plain management, environmental permitting, and other programs. Through effective system design, information can be exchanged across governments in a manner that honors the Tribe's sovereignty while supporting its overall sustainable land and natural resource management goals. EBCI's enterprise IGIS is benefiting the Tribe and neighboring governments by dramatically reducing the time it takes to gather information and make fast and accurate decisions. In many cases, decisions that once took days, weeks, and months can now be made in a matter of minutes or seconds. The IGIS serves as an example of the next generation of enterprise systems designed to streamline business practices and provide efficient governmental services by tribal governments to all of their stakeholders and beneficiaries. The EBCI is using GIS as the core component of improving geographical related workflow operations in the services-driven economy.

Enterprise GIS at the Chickasaw Nation

John Ellis, Director, Department of Geospatial Information, the Chickasaw Nation
Ada, Oklahoma

Through other chapters in this book, it is clear that the advanced GIS platform at the Chickasaw Nation is aiding in the Tribe's governance and provision of services to its members. The enterprise GIS program at the Chickasaw Nation is managed by the Department of GeoSpatial Information (GSI) within the Division of Housing and Tribal Development; however, it supports even more tribal government programs than discussed elsewhere in this book.

Providing citizen services across 7,443 square miles in South Central Oklahoma, covering all or parts of thirteen counties, is not trivial. In support of government programs, the Tribe began using GIS in 1997. At that time there were several divisions within the Chickasaw Nation that handled a variety of aspects of any given government project. Over a nine-year period, the number of Chickasaw Nation employees grew from under 2,000 to more than 11,500. Because of this increase, the knowledge about projects became fractured into several locations. Tribal boundaries, tribal lands, and allotted lands were frequently under scrutiny. The GSI department, by implementing an enterprise GIS, offered Chickasaw decision makers a way to create one central repository of data for these parcels that would ensure consistency of information within the entire Chickasaw Nation.

Several factors related to data management and data dissemination made evident the need for an enterprise GIS system. Questions were often asked about responsibility for maintaining certain lands and facilities, storage for relevant documentation, and existence of certain data. The large increase in employee numbers also meant an exponentially large increase in tribal data. Data was stored on individual computers, shared folders, CDs, jump drives, and even hard copy. This increased the risk of data corruption and misplacement. Additionally, map requests took considerable time to prepare due to the lengthy data acquisition and editing stages. Prior to the implementation of the enterprise GIS within the Chickasaw Nation, there were several GIS workstations that shared a single storage system located on a centralized server within the data center. The server was used as file storage, and all applications resided on the workstations. This configuration lacked the scalability to meet the needs of the tribal government.

The first steps toward the enterprise system focused on defining user requirements. The System Development Life Cycle (SDLC) design model was used and performed information gathering within existing departments that had conducted business with the GSI department. In addition, several meetings were conducted with department personnel to determine expectations of an enterprise GIS system and to ensure all issues were discovered and addressed. The GSI department established a scope, timeline, and budget for the project, along with initial milestones to be presented as individual steps in completing the system.

The current enterprise design was accomplished after many discussions and on-site visits from Esri support teams. Using information about utilization, trends, and expected growth, a model was developed using network standards provided by the IT department. These standards provided the manufacture and models by which all servers are currently incorporated into the network. The department used these standards and the recommendations from Esri to construct a working model of the GIS server farm. This model was submitted to the IT department for approval. Prior to approval of the network design, a system administrator position had to be filled to support the server farm, storage area network, and backup and recovery requirements for the project.

Products and services that have resulted from the enterprise GIS include web services, integration of Image Server, routing, and map book functionality, a Pictometry toolbar, and a direct link to the Lighthorse Police Department's CAD system. Direct beneficiaries of the enterprise GIS are the governor's office; the Chickasaw Nation Legislature; and the tribal housing and development, roads, realty, cultural preservation, commerce and Lighthorse Police Departments. Eighty-eight percent of the divisions of the Chickasaw Nation are beneficiaries of the enterprise GIS through maps and reports. Tribal government leadership benefits from GIS services provided by having access to data that allows for real-time, informed decisions. In addition, several entities outside of the Chickasaw Nation, including East Central University (contract work and hiring students for internships), local law enforcement, fire departments, and the BIA, have received assistance in the form of maps or reports.

GIS personnel within the GeoSpatial Information Department who were instrumental in developing the enterprise GIS include the director, GIS manager, GIS systems administrator, web applications developer, and GIS analyst. The IT personnel who were instrumental in developing the enterprise GIS included the IT director, IT project manager, and a database administrator.

The Chickasaw enterprise GIS platform operates on five IBM Blade Servers with one H Chassis capable of holding up to fourteen blade servers. The GSI enterprise platform was designed to include a testing environment that is an exact replica of the production environment. The department is running on ArcGIS Server and Image Server, and one 12 TB San Server that stores approximately 5.12 TBs of geographic data. GSI has created a comprehensive geographic information system with large amounts of data. The enterprise system currently contains six geodatabases with approximately 3,500 GBs of geographic data and 2.84 TBs of imagery with approximately 15 GBs of that imagery processed through two image servers at this time. The enterprise GIS includes standard operating procedures to capture, download, verify, and differentially correct GPS data before it is input into the enterprise geodatabase. The data is reviewed for quality assurance/quality control (QA/QC) and then posted to the default version in the geodatabase. Remotely sensed imagery is input into the Image Server by the GIS system administrator through a consistent standard operating procedure.

The GIS enterprise network implementation was started from scratch, therefore there was little risk involved during the actual implementation. Precautions were taken to ensure the existing data located on individual workstations and the file storage servers were backed up to tape prior to starting the data migration portion of the project. Other precautions included backups of all software prior to installation on the servers to prevent loss during migration. Extensive steps were also taken to ensure all software was compatible, not only with the other software but also the hardware.

Feasibility studies were conducted to ensure the new GIS network would not overload the bandwidth of the existing network and new switches were installed in the local IT closet. All installs of hardware and software were performed at night to ensure minimal network disruptions. Bandwidth testing was performed at various stages during the implementation. The initial enterprise GIS was tested in real-time, but a fully-functioning testing environment is now used for all service patch and upgrade installations.

Numerous challenges were met to provide the Chickasaw Nation with an enterprise GIS. Challenges included limited knowledge by staff members about enterprise GIS systems, acquiring budget approval, interdepartmental cooperation, and convincing Chickasaw Nation decision makers that GIS was an effective and efficient decision-making tool. These challenges were overcome by gaining education and awareness through Esri classes, conferences, and consultations with outside parties; including IT staff in GIS training so they could offer support for the system; presenting GIS to administrators and directors with real-world Chickasaw Nation scenarios; involving and educating other departments about GIS through needs assessment questionnaires; and being patient and giving the project time to come together.

ArcGIS Server is used to reach a larger audience, and there are currently dedicated intranet websites for the Chickasaw Nation Legislature, and the tribal housing and development, commerce, communications, maintenance, cultural preservation, commerce and Lighthorse Police Departments. Other projects are in progress, with plans to provide the Chickasaw citizens with an external web application in the near future. As a result of implementing an enterprise GIS, the Chickasaw Nation now has a central repository for all tribal spatial data with appropriate permissions and security measures.

The implementation of enterprise GIS has allowed Chickasaw decision makers to more efficiently and effectively manage tribal holdings, from purchase to assigning use, to tracking progress of trust status. Unforeseen benefits of this project include a better working relationship with other departments, quicker turnaround of GIS products, satisfied customers, increased GIS knowledge, increased personnel, state-of-the-art technology, and a larger operating budget. Information is readily available to employees, and GIS production takes a fraction of the time it did previously.